AF581798

Selected Papers from MEDPOWER 2018

Selected Papers from MEDPOWER 2018

The 11th Mediterranean Conference on Power Generation, Transmission, Distribution and Energy Conversion

Editors

Igor Kuzle
Tomislav Capuder
Hrvoje Pandžić

MDPI • Basel • Beijing • Wuhan • Barcelona • Belgrade • Manchester • Tokyo • Cluj • Tianjin

Editors

Igor Kuzle
University of Zagreb
Croatia

Tomislav Capuder
University of Zagreb
Croatia

Hrvoje Pandžić
University of Zagreb
Croatia

Editorial Office
MDPI
St. Alban-Anlage 66
4052 Basel, Switzerland

This is a reprint of articles from the Special Issue published online in the open access journal *Energies* (ISSN 1996-1073) (available at: https://www.mdpi.com/journal/energies/special_issues/MEDPOWER_2018).

For citation purposes, cite each article independently as indicated on the article page online and as indicated below:

LastName, A.A.; LastName, B.B.; LastName, C.C. Article Title. *Journal Name* **Year**, *Volume Number*, Page Range.

ISBN 978-3-0365-0772-9 (Hbk)
ISBN 978-3-0365-0773-6 (PDF)

Contents

About the Editors

Tomislav Capuder is an Associate Professor at the University of Zagreb Faculty of Electrical Engineering and Computing. He has received multiple awards for his research and scientific contributions (Young Danube Scientist Award, Vera Johannides Award by Croatian Academy of Technical Sciences etc.). Prof.Capuder has published over 100 journal and international conference papers and has participated in over 120 industry sponsored projects related to various topics of power system planning and operation. Currently, he coordinates or participates in several EU and nationally funded projects. His areas of expertise include: energy systems planning and modelling, integrated infrastructures, distributed energy systems, energy markets, environmental issues in power system. Tomislav is acting as a member of the IEEE Energy Policy Work Group and a member of Croatian CIGRE and SDEWES as well as a member of international CIRED work group on multi-energy systems, energy storage and distribution network resilience.

Hrvoje Pandžić is an Associate Professor and Head of the Department of Energy and Power Systems at the University of Zagreb Faculty of Electrical Engineering and Computing, Croatia. He is a Project Leader at ICENT's Institute for Energy. From 2012 to 2014, he was a Postdoctoral Researcher with the University of Washington, Seattle, WA, USA. He coordinated H2020 ERA Net Smart Grids + project uGRIP, as well as two large national research projects. He is an author of over 40 Q1/Q2 journal research papers and he led or took part in over 120 technical studies for various clients. He is a recipient of the Award Science by the Government of the Republic of Croatia for 2018 and the Award for the highest scientific and artistic achievements in Croatia by the Croatian Academy of Science and Arts for 2018. His research is focused on integration of renewable energy sources through energy storage smart grids.

Igor Kuzle is a Full Professor at the Department of Energy and Power Systems at the University of Zagreb Faculty of Electrical Engineering and Computing. He serves in 10 international journals as an associate editor or a member of editorial board. Igor Kuzle published three books and more than 300 journal and conference papers including technical studies for utilities and private companies. He was the project leader for more than 60 technical projects for industry. Currently, he is the principle investigator in five research projects related to increasing of flexibility of the power system to facilitate integration of RES and electric vehicles. Prof. Kuzle was awarded the National Science Award for 2017 for his contribution to science in the field of advanced networks in the transmission system and the FER Science Award for his outstanding research contribution in the period from 2010 to 2015. He received the HAZU Award for 2019 for his scientific contribution from the application of various concepts of managing advanced electricity networks in order to increase the flexibility of the electricity system and enable the mass integration of renewable energy sources. His full publication list can be found at http://bib.irb.hr/lista-radova?autor=206894.

Preface to "Selected Papers from MEDPOWER 2018"

This book is a collection of the best papers from the 11th Mediterranean Conference on Power Generation, Transmission, Distribution and Energy Conversion (Medpower 2018) conference, held in Dubrovnik in November 2018. The papers selected to be published in the Special Issue of *Energies* reflect the wide range of topics covered at the 11th Medpower conference, where experts from all over the world presented their research emphasizing and focusing on practical engineering solutions that need to be developed to enable the transition to a clean/decarbonized and sustainable energy future. Most of the researchers and engineers would agree that the largest transformation will be occurring on the distribution level, by the end consumers. This changes how distribution system operators (DSOs) plan their network, as the conventional approach of system reinforcements can be substituted by using the flexibility of prosumers and ensuring that the constraints imposed by the regulatory framework guaranteeing the security of supply and quality of power are maintained. To capture all these aspects and possibilities, DSOs are designing network tariffs to stimulate this complex process; an interesting approach on this is described in a selected paper: www.mdpi.com/1996-1073/12/7/1203. Different incentives or control actions are needed in the operational framework, especially with the change brought by power electronics interfacing low carbon technologies to the known operational principles. A proposal of a new control mechanism for mitigation of negative sequence currents under asymmetrical voltages in the low-voltage distribution network is presented: https://www.mdpi.com/1996-1073/12/7/1362. The proposed real-time control involves the use of the day-ahead (or longer) schedules based on quality forecasting of uncertainties. Predicting uncertain behavior is a challenge, especially with new technologies such as electric vehicles (EVs), where a limited amount of real-world data are available; the work in www.mdpi.com/1996-1073/12/7/1341 proposes a method for forecasting EV behavior and profiles and assesses the impact on future distribution networks. The evolution on the distribution grid side not only creates challenges in dealing with two-way energy flows, but also creates opportunities for active participation and reducing the dependency on the upstream system and increases system stability (www.mdpi.com/1996-1073/12/13/2470). Although fewer changes may occur on the transmission side, the main task of maintining the stability, security, and reliability of the power system is becoming more complex for transmission system operators (TSOs) as the nodes in the network, which traditionally used passive consumption, are now active participants and injectors of both active and reactive power. One of the inexpensive remedial actions and tools to mitigate congestions, both current and voltage, caused by this change is phase shifters, as described in www.mdpi.com/1996-1073/12/11/2167. To prevent congestion in the transmission system, which can result in significantly higher market prices in one zone, the technologies alleviating congestion need to be supported by both market models and mechanisms (https://www.mdpi.com/1996-1073/12/3/490) as well as by increased real-time observability. However, connectivity and digitalization may be accompanied by the low data availability and the risk of cyber-attacks, described in www.mdpi.com/1996-1073/12/13/2598. Finally, the transition to a low-carbon future cannot be achieved only by small-scale production and consumption activities in the distribution networks. Bulk generation will remain the backbone of the power system for decades to come, both in existing and the new and renewable power plants. Their secure operation is the key for system stability and reliability; however, their outage can have additional impacts on the environment if not managed properly. Both a model and practical test from a plant outage in 2016 of the real-world nuclear power plant, Krško, are presented in www.mdpi.com/1996-1073/12/11/2176. The aspects of power system reliability should not be observed in the operational stage, or in case of contingency, but they need to be integrated in the planning aspects as well (www.mdpi.com/1996-1073/12/9/1714). The approach to planning of generation units needs to be multidimensional and cover different influencing factors such as technical specifications, geological characteristics, spatial data, energy and heat prices, and social and environmental impact (www.mdpi.com/1996-1073/12/9/1597). The Guest Editorial Board would like to thank *Energies* for providing the opportunity to organize this Special Issue, the authors for their innovative and valuable contributions, and the reviewers for their prompt and comprehensive feedback and suggestions. Guest Editorial board Tomislav Capuder, Igor Kuzle, and Hrvoje Pandžić acknowledge the support of the Croatian Science foundation through their financing of the projects IMAGINE, FLEXIBASE, and WINDLIPS.

Igor Kuzle, Tomislav Capuder, Hrvoje Pandžić
Editors

 *energies*

Article

Evaluating the Evolution of Distribution Networks under Different Regulatory Frameworks with Multi-Agent Modelling

Miguel Manuel de Villena [1,*], Raphael Fonteneau [1], Axel Gautier [2] and Damien Ernst [1]

1 Montefiore Institute, Department of Electrical Engineering and Computer Science, University of Liege, 4000 Liege, Belgium; raphael.fonteneau@uliege.be (R.F.); dernst@uliege.be (D.E.)

2 HEC Liege, LCII, University of Liege, 4000 Liege, Belgium; agautier@uliege.be

* Correspondence: mvillena@uliege.be

Received: 5 March 2019; Accepted: 26 March 2019; Published: 28 March 2019

Abstract: In the context of increasing decentralised electricity generation, this paper evaluates the effect of different regulatory frameworks on the evolution of distribution networks. This problem is addressed by means of agent based modelling in which the interactions between the agents of a distribution network and an environment are described. The consumers and the distribution system operator are the agents, which act in an environment that is composed by a set of rules. For a given environment, we can simulate the evolution of the distribution network by computing the actions of the agents at every time step of a discrete time dynamical system. We assume the electricity consumers are rational agents that may deploy distributed energy installations. The deployment of such installations may alter the remuneration mechanism of the distribution system operator. By modelling this mechanism, we may compute the evolution of the electricity distribution tariff in response to the deployment of distributed generation.

Keywords: multi-agent modelling; distribution networks; distributed generation integration; power system economics; regulation; optimisation

---

1. Introduction

One of the primary enablers of the energy transition is the widespread growth in the integration of distributed energy resources (DER) into the electricity mix [1]. For this reason, distributed generating technologies as, for example, solar photovoltaic (PV), have been (and are being) globally stimulated by means of policies and directives in order to foster their deployment (see for instance the European Parliament Directive 2009/28/EC [2]). These policies are usually translated into different incentive mechanisms, such as feed-in tariffs (FiT), feed-in premiums (FiP), or other monetary aids which help improve the business models of DER as generating technologies. Along with the incentive mechanisms, there are several indirect key drivers of DER deployment. Two such drivers are the distribution tariff design (which for simplicity will be called tariff design in this paper), and the technology costs. Regarding the former, they are typically regulated by the incumbent regulatory authority, which can be regional (e.g., in Belgium the tariffs are regulated by three different regulatory authorities depending on the region, namely, Brussels, Flanders, and Wallonia) or national (e.g., in Germany the tariff design is regulated at a national level). As for the technology costs, over the last few years, these have been decreasing, and, according to the projections, they may still progressively decrease during the coming decade, owing to economy of scales and technological maturity [3]. All these factors combined and, in particular, the widespread use of incentive mechanisms, have contributed to a substantial deployment of DER; however, such a DER integration might conceal the potential to create both technical problems (e.g., under- and over-voltages) [4] and regulatory challenges (e.g., cross-subsidisation amongst electricity consumers) [5–7].

These regulatory challenges are multifaceted, and notably comprise, amongst others: (i) equity problems derived from an unfair allocation of the electricity distribution costs, which may lead to cross-subsidies amongst the consumers of the distribution networks [6]; (ii) the potential failure of the distribution system operators (DSOs) remuneration mechanisms [5]; or (iii) a generalised increase in the distribution tariff [8], i.e., the distribution component of the overall retail electricity price, the price end consumers are exposed to, which is composed of energy costs, transmission costs, distribution costs, taxes, and the retailer margin.

The scope of this paper is to quantitatively assess the nature and extent of these regulatory challenges, making use of a simulation environment to evaluate how the deployment of substantial amounts of DER may alter the remuneration mechanisms of DSOs and how this, in turn, may have an impact on the distribution tariffs. In particular, we introduce the first elements of a methodology to compute the impact of different regulatory frameworks on the agents of a distribution network. This methodology allows for dynamically evaluating such impacts, moving beyond the static assessments which are usually performed. In a static analysis, the variables of the system (e.g., deployed DER or technology costs) are computed once (i.e., DER are deployed responding to some fixed action from the regulator). In a dynamical system approach, each variable evolves over time, rendering different states of the system at every evaluated time-step (i.e., the reaction of DER to the regulator is evaluated at several points in time). In this context, the complete evolution of the system can be computed by fixing the set of rules (i.e., the regulatory framework) controlling the interactions between the variables. The regulatory framework describes the distribution tariff design, the remuneration mechanism of the DSO, the incentive mechanisms, and the technology costs. Finally, this methodology enables employing different regulatory frameworks, allowing for testing the short to middle run effects of distinct regulatory practices on the distribution networks and their agents.

For the remainder of the paper, Section 2 documents previous works dealing with the regulatory challenges posed by a large integration of DER. Section 3 introduces a high level description of the simulator. Section 4 explains how the regulatory framework (composed of tariff design and incentive mechanism) is modelled. Section 5 exhibits a case study in which we make use of the developed simulator. Finally, Section 6 concludes and exposes the limitations of our approach.

2. Related Works

Studying the regulatory challenges existing in distribution networks has been the subject of debate over the last few decades, as the available literature reveals. In one of the first research papers on this topic [9], the authors identify two main elements to regulate: setting the distribution tariff allocating the total costs among all the users, and establishing an adequate remuneration mechanism for the DSO. Moreover, they propose a remuneration mechanism based on a revenue limitation scheme, as previously described in [10]. The two first documents dealing with the issue of distributed generation (DG) in distribution networks are [11,12]. The former focuses on the impact of DG on the power systems, while the latter discusses the effects of regulation on the deployment of DG. The concept of DG as generating technologies, generally of reduced installed capacity, and connected to the distribution networks is introduced in [13], where the authors showcase different DG technologies and their different costs. As mentioned in the Introduction, the foremost drivers of DG integration (in which DER are included) are identified. Two of them are the distribution tariff design on the one hand, and the use of incentive mechanisms on the other hand. The existing literature can be divided accordingly.

2.1. Distribution Tariff Design

Concerning distribution tariff design, most of the existing literature focuses on exploring how distribution costs should be charged to end consumers. A series of rules for the design of distribution tariffs can be found in [14], as well as in the CEER report [15]. According to these works, the design of a tariff should account for the choice of remuneration mechanisms, the tariff structure, and the allocation of allowed costs. Furthermore, the key regulatory principles to design a tariff are identified, e.g., sustainability, non-discriminatory access, efficiency, transparency, or tariff additivity. These principles are, by and large, shared in [16,17], where relevant regulatory principles are listed. In [18], the authors recommend that DG (both DER and combined heat and power) pay regulated shallow connection costs in order to facilitate the integration of these generation resources. The discussion shallow vis-à-vis deep connection costs is also addressed in [7,19–23], where diverse methodologies are assessed. In short, deep connection costs comprise the connection cost itself as well as the costs derived from reinforcing the network, and shallow connection costs consist only of the connection cost, whereas the potential costs of reinforcing the network are socialised via the distribution tariff. Some of the existing works experiment with different distribution tariff designs, looking into their impact on DG and on the DSO ability to recover its costs. In this regard, the authors in [24] explore designs based on the cost-causality principle, claiming that such tariffs function better than average cost distribution tariffs to recover fixed network costs. In [16], the authors suggest a method to assign costs according to the same principle, based on peak demand, overall energy demand, and geographical location. Moreover, in this work, it is highlighted that, since consumers may react to the tariffs, setting an adequate tariff might be an iterative process. In [25], the researchers propose a way of taking into consideration the impact of DG on the cost-causality criterion used to design and compute distribution tariffs. In these studies, different technologies are mentioned for measuring the energy consumption and production of the DER installation, namely net-metering and net-billing:

- Net-metering (NM): consists of one meter that records imports (DER $\leftarrow$ Grid) by running forwards, and exports (DER $\rightarrow$ Grid) by running backwards. Therefore, both directions are assigned with the same monetary value, namely the retail electricity tariff. Additionally, if the exports exceed the imports, the excess is not remunerated.
- Net-billing (NB) also called net purchase and sale: consists of two independent meters for imports and exports; in this setting, imports are charged at retail price, and exports are compensated at a selling price. There is, in principle, no limit to the amount of exports allowed.

Several authors have discussed the impact of these two systems. In [26], a model to evaluate the impact of NM policies in introduced, concluding that this system is extremely beneficial for consumers owners of a DER installation (prosumers) but may create macroeconomic problems such as the increase of the distribution tariff. Similar analyses are conducted in [5,27], where the authors compare NM with NB, claiming that NM may lead to both cross-subsidies amongst the users of a distribution network and an uncontrolled increase in distribution prices, also known as the death spiral of the utility [8,27]. Analogous conclusions are drawn by [6], where the authors state that NM presents a trade-off between incentivising DG and securing the financial stability of the DSO. In [28,29], NM in the United States is analysed; these papers suggest that NM enhances the value if behind-the-meter; additionally, they claim that the potential feedback created by NM (i.e., the utility death spiral) is rather modest.

Another way of spurring the deployment of DER installations is by introducing changes in the method used to charge consumers and prosumers for their electricity consumption. Various methods have been explored in all the previous works, e.g., capacity or demand tariffs (€/kW), volumetric tariffs (€/kWh), fixed tariffs (€/connection), or time-of-use (ToU) tariffs. In this regard, the analysis in [30] shows that, when applying volumetric distribution charges, in a setting where NM is in place, an increase in the distribution tariff leads to household PV deployment. In [31], the author demonstrates that a peak demand capacity tariff is more efficient and cost-reflective than its volumetric counterpart.

2.2. Incentive Mechanisms

Concerning incentive mechanisms (or support schemes), several authors have examined the effect of FiTs. In [32], FiTs are compared with traditional schemes such as renewable obligations, proposing a two-part FiT with capacity and energy payments which the authors claim to be a more effective framework for fostering the deployment of DER than the existing alternatives. The authors in [33] review the regulatory and policy frameworks of FiT schemes, laying stress on how these have affected the solar PV market. They highlight that, due to generous tariffs, the market bloomed in 2008; nevertheless, FiTs have failed to continue supporting PV integration since they tend to distort the electricity prices leading to economic instability. On the same topic, Ref. [34] shows that FiTs are likely to work better than quantity-based systems (e.g., quota-obligation) when it comes to fostering DER.

In addition, a few works can be found assessing the use of incentive mechanisms to promote the deployment of DER, for a range of different tariff designs. For example, in [35,36], the authors analyse the use of FiTs in combination with NM and with NB. However, the results of these studies are inconclusive insofar as they greatly depend on the initial conditions (e.g., level of DER penetration or distribution prices).

2.3. Modelling

To date, the level of modelling in all these analyses is rather limited owing to the complexity of representing abstract regulatory principles in an exact manner. Furthermore, modelling the behaviour of prosumers is complex since they may not act rationally (see for instance [37]). For these reasons, in most of the existing literature, the penetration of DER as well as the distribution prices are considered parameters to study with little or no interaction between them. There are some works, nonetheless, where this is addressed. In [38], the authors highlight the importance of designing efficient distribution charges in the context of increasing DER integration, claiming that the network peak is the main driver of network investment. A model is introduced in this paper in which users can react to distribution charges by deploying fix-sized DER installations in order to overcome high distribution charges. Moreover, in [39], a model of interaction between prosumers and DSO is proposed comparing NM with NB; in this model, prosumers react to distribution prices by deploying optimally sized DER installations, the tariff is then updated by the DSO, responding to a change in energy consumption. In [40], a model including capacity charges and injection fees is proposed, concluding that transitioning to rate structures including capacity charges will not disrupt the adoption of PV and will lower the costs of consumers. Finally, in [41], a game-theoretical model is proposed to assess volumetric and capacity tariffs, their impact on the potential prosumers, and the consequences for the consumers.

2.4. Motivation

As we can see, some of the questions proposed in our paper have been to some extent studied in the previous literature, although from a purely qualitative standpoint. Only a few works exist tackling this issue from a more quantitative perspective, using mathematical tools to simulate the behaviour of end-users in a distribution network and, although with limitations, to estimate the repercussions of such behaviours for the distribution networks and, in particular, for the distribution tariff. Our research paper stands on the shoulders of all these works, proposing a methodology to quantify the development of distribution networks across time, as a function of the DER deployment and the distribution tariff evolution. Furthermore, an interaction between DER deployment and distribution rates is modelled by which they impact one another and evolve over time, attaining—or not—an equilibrium after the simulation is completed (the horizon is reached). Our work includes the analysis of distribution tariffs based on different proportions of volumetric and fixed fees, as well as different incentive mechanisms (NM and NB), in a simulation environment in which the actors are residential consumers some of which may become prosumers, and the DSO.

3. Simulator

The simulator introduced in this section relies on multi-agent modelling. It allows modelling every consumer/prosumer of the distribution network as a rational agent, who may deploy an optimally sized DER installation if it is cost-efficient compared to the retail prices. Furthermore, the DSO is also modelled as an agent that can adjust the distribution tariff to recover its costs of providing the distribution service. To assess the evolution of the distribution network, we introduce a discrete time dynamical system that enables computing the actions of the agents at every time step. Finally, to compare different regulatory frameworks, we introduce the concept of environment, which includes all of the rules characterising them. In our simulator, the agents interact (perform actions) within a particular environment. By modifying the environment, we also modify the actions of the agents, allowing the assessment of the distribution network evolution under different regulatory frameworks.

3.1. Environment Representation

Every environment is built with three distinct elements: (i) tariff design, (ii) incentive mechanism, and (iii) technology costs (assumed linearly decreasing over time).

We introduce distinct tariffs based on different proportions of volumetric fees, paid in €/kWh, and fixed fees, paid in €/connection. Furthermore, we include two different incentive mechanisms for the consumers to deploy DER, NM and NB, which have previously been explained.

3.2. Actions of the Agents

There are two types of agents:

- Consumers: at the beginning of the simulation, they simply draw electricity from the distribution network. However, as the simulation proceeds over the discrete time dynamical system, they take actions to gradually deploy optimally sized DER installations, becoming prosumers. The prosumers may draw (import) and inject (export) electricity to the distribution network. To model the planning and operation of these agents, i.e., the computation of their electricity trades (imports and exports), and the transition consumer $\rightarrow$ prosumer (DER deployment), we resort to an optimisation framework instantiated as a mixed integer linear program (MILP). This MILP is loosely based on the LP found in [42], and aims at minimising the levelized cost of electricity (LCOE) of the DER installation. The potential investment allowed for the consumers consists of an optimally sized PV installation with or without batteries (depending on the optimisation).
- DSO: the actions of this agent consist in adjusting the distribution tariff according to the environment in place. For example, after collecting revenues, this agent may increase or decrease the distribution tariff, under the assumption that it must break-even (here it is assumed that if the DSO collects the total amount of allowed revenues, it will completely cover its costs). The DSO cannot modify the tariff design, since this is imposed by the environment. Hence, if the tariff design set by the environment consists of a fully volumetric distribution tariff, the DSO will be able to adjust the price per kWh, but it will not be able to recover costs by applying extra charges to the distribution network consumers. The operation of this agent is modelled with its remuneration mechanism.

3.3. Discrete Time Dynamical System

The actions of the agents lead to the evolution of the distribution network. The consumers, by deploying DER, reduce their dependency on the distribution network, lowering their apparent consumption, which refers to the energy recorded by the meter at the end of the billing period. In response to the consumers actions, the DSO will adjust the distribution tariff according to its remuneration mechanism. Thus, we can compute the distribution network evolution as a function of the agents actions, by evaluating them at every time step of a discrete time dynamical system.

Let $n \in \mathcal{N}$ denote the time index of the discrete time dynamical system, with $\mathcal{N} = \{0, \ldots, N-1\}$. At every time step n, our simulator computes the actions of the agents, controlling the transition from n to $n+1$. This computation follows a specific order: (1) the transition consumer $\rightarrow$ prosumer is calculated, determining their apparent consumption Ξ_n; (2) the DSO adjusts the distribution tariff $\Pi_n^{(dis)}$. In Figure 1, a time-line of the discrete time dynamical system can be found.

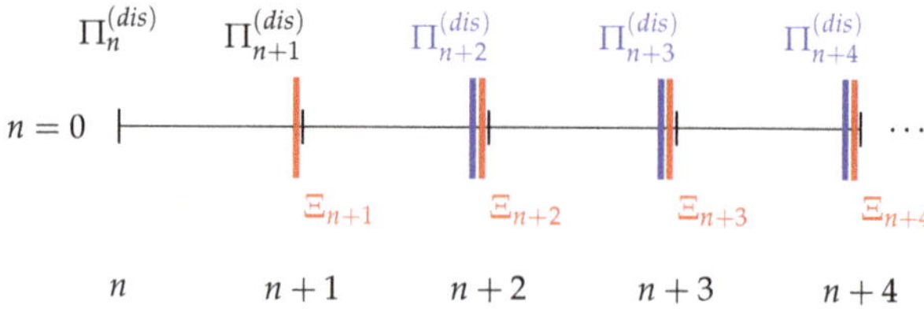

Figure 1. Time-line of the discrete time dynamical system. The simulation starts by assuming a distribution tariff $\Pi_n^{(dis)}$. Then, at every time step, there is a transition consumer $\rightarrow$ prosumer leading to a change in the aggregated apparent consumption Ξ_n. This change induces an adjustment of the distribution tariff $\Pi_n^{(dis)}$.

The first billing period is necessary so that the consumers can observe their electricity bill under the initial conditions. Then, the transition consumer $\rightarrow$ prosumer can be computed, and, from it, we determine the total apparent consumption Ξ_{n+1} (which corresponds to the period $n+1$ to $n+2$). Since the consumption during the period n to $n+1$ and the consumption under the initial conditions are the same, the distribution tariff does not change ($\Pi_n^{(dis)} \equiv \Pi_{n+1}^{(dis)}$). However, once Ξ_{n+1} is computed, it induces a change in the distribution tariff for the following period $\Pi_{n+2}^{(dis)}$ (after the DSO observation of its revenue during the period $n+1$ to $n+2$). We assume that the consumers can react immediately to this distribution tariff adjustment since they already have knowledge regarding their consumption. Then, the aggregated apparent consumption Ξ_{n+2} can be calculated. The discrete time dynamical system continues in this fashion until no more consumers can turn into prosumers, or until the stopping criteria are met: when reaching the finite time horizon N, or when the DER are not economically profitable.

Every time step of the discrete time dynamical system, except for the first one, is computed with one run of our simulator. Thus, the developed simulator is run recursively to simulate the complete dynamical system. The end of one run will be used as starting point for the next one. The flow diagram representing an outline of one run of the simulator can be found in Figure 2.

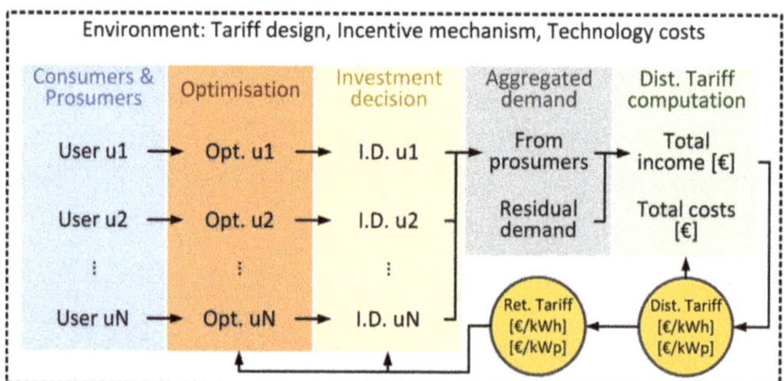

Figure 2. Flow diagram of the proposed multi-agent simulator. The flow of actions occurs from left to right. The distribution network consumers undergo individual MILP optimisations to minimise their LCOEs. A transition consumer $\rightarrow$ prosumer is computed (investment decision tab (yellow) on the Figure), and finally the DSO adjusts the distribution tariff.

The simulation starts with a pool of consumers who may become prosumers at any point of our discrete time dynamical system. These agents, characterised by their load, are modelled through an MILP to plan and operate a DER installation minimising their LCOEs. Such an optimisation requires the retail electricity price at each time step, as well as the demand profile of the consumers. After the optimisation, a transition consumer $\rightarrow$ prosumer is computed by comparing the costs of the consumers with and without DER installation. The aggregated apparent consumption Ξ_n is then calculated and added to the residual demand of the system (those consumers of the distribution network who cannot deploy DER due to technical or economic constraints). Finally, the DSO revenues are computed, and, assuming constant costs across the discrete time dynamical system, the distribution tariff $\Pi_n^{(dis)}$ for the following time step is determined so as to fully recover those costs.

4. Modelling the Regulatory Framework

In this simulator, we introduce the concept of environment as the container of the set of rules that characterise the distribution network, namely the tariff design, the incentive mechanism, and the technology costs. Hence, to model the agents actions, we must take into account the distinct possible environments. Every single agent takes individual actions; therefore, we need to introduce the set $\mathcal{I} = \{1, \ldots, I\}$ representing the consumers/prosumers, where I is the number of consumers/prosumers. In the following, we present the differences in the simulator, depending on the tariff design and the incentive mechanism.

4.1. *Tariff Design*

We may use a fully volumetric design, or a two-part design with a variable (volumetric) and a fixed (per connection) term.

4.1.1. Fully Volumetric

Under this setting, the individual electricity costs $\psi_{i,n}$ of the agents in I are calculated as follows:

$$\forall i, n \in \mathcal{I} \times \mathcal{N} \quad \psi_{i,n} = \xi_{i,n} \cdot \left(\Pi_n^{(dis)} + \Pi^{(en)} \right), \tag{1}$$

where $\xi_{i,n}$ represents the individual apparent consumption of the ith prosumer at the nth time step, $\Pi_n^{(dis)}$ is the distribution tariff set by the DSO, and $\Pi^{(en)}$ represents the costs of energy, transmission and taxes, held constant across the simulation.

The DSO revenues are calculated as:

$$\forall n \in \mathcal{N} \quad R_n = \Pi_n^{(dis)} \cdot (\Omega + \Xi_n), \tag{2}$$

with Ω being the residual demand of the system (held constant), and Ξ_n is the aggregated apparent consumption of the consumers/prosumers, which is calculated as $\Xi_n = \sum_{i=1}^{I} \xi_{i,n}$

4.1.2. Two Part

In this case, we introduce a fixed term in the calculations. The electricity costs of the consumers/prosumers are calculated as follows:

$$\forall i, n \in \mathcal{I} \times \mathcal{N} \quad \psi_{i,n} = \xi_{i,n} \cdot \left(\Pi_n^{(dis)} + \Pi^{(en)} \right) + c_i, \tag{3}$$

where the term c_i is set at the beginning of the simulation (see Equation (8)) and kept constant. As for the DSO, its revenues are computed as follows:

$$\forall n \in \mathcal{N} \quad R_n = \Pi_n^{(dis)} \cdot (\Omega + \Xi_n) + C, \tag{4}$$

where $C = \sum_{i=1}^{(I+J)} c_i$, with J being the amount of consumers who make up the residual demand Ω.

4.2. Incentive Mechanism

We may use NM and NB. This choice impacts the individual apparent consumption, and, as such, the aggregated one.

4.2.1. NM

The individual apparent consumption of the consumers/prosumers is given by:

$$\forall i,n \in \mathcal{I} \times \mathcal{N} \quad \xi_{i,n} = \max\left\{0, \left(\rho_{i,n}^{(-)} - \rho_{i,n}^{(+)}\right)\right\}, \tag{5}$$

where $\rho_{i,n}^{(-)}$ and $\rho_{i,n}^{(+)}$ are, respectively, the imports and exports of the ith prosumer at the nth time step.

4.2.2. NB

In this case, the exports do not affect the apparent consumption, thus:

$$\forall i,n \in \mathcal{I} \times \mathcal{N} \quad \xi_{i,n} = \rho_{i,n}^{(-)}. \tag{6}$$

4.3. Distribution Tariff Update

For every option of tariff design and incentive mechanism, the distribution tariff is updated at every time-step according to the following equation:

$$\forall n \in \mathcal{N} \quad \Pi_{n+1}^{(dis)} = \frac{\Theta + \Delta_n - C}{\Omega + \Xi_n}, \tag{7}$$

where Θ are the costs of the DSO, which are calculated as the revenues of the first time step R_0, and held constant across the dynamical system. The imbalance from the previous period is introduced with the difference $\Delta_n = \Theta - R_n$. In other words, Θ represents the costs of the DSO, Δ_n represents the "missing money" from previous periods, and C represents the money recovered through fix charges, and the sum of these parameters thus represents a quantity in €. Then, Ω represents the residual demand of the system (only consumers), Ξ_n represents the demand of the prosumers, and the sum of these parameters is therefore a quantity in kWh. The mechanism works in a way the tariff $\Pi_{n+1}^{(dis)}$ is updated according to costs divided by demand.

5. Case Study

To assess the impact of different environments on the distribution network evolution, we introduce a case study in which we compare nine different environments (regulatory frameworks). The simulator necessitates a set of consumers/prosumers to work. Each consumer/prosumer is characterised by a demand profile and a production profile. Once we have produced the set of consumers/prosumers, we evaluate: (i) four distinct designs with decreasing proportions of costs recovered through volumetric charges being replaced by fixed ones (where the incentive mechanism is kept fixed for all of them), and (ii) five different incentive mechanisms (where the proportion of volumetric charges is kept fixed for all of them). The different assessed environments are presented next.

- Different proportions of volumetric and fixed charges:
  - E1: environment with 100% volumetric charges. NB is used as incentive mechanism with a selling price of 4c€.
  - E2: environment with 75% volumetric charges and 25% fixed charges. NB is used as incentive mechanism with a selling price of 4c€.

- E3: environment with 50% volumetric charges and 50% fixed charges. NB is used as incentive mechanism with a selling price of 4c€.
- E4: environment with 25% volumetric charges and 75% fixed charges. NB is used as incentive mechanism with a selling price of 4c€.

- Different incentive mechanisms:
  - E5: environment with 100% volumetric charges. NM is used as an incentive mechanism.
  - E6: environment with 100% volumetric charges. NB is used as incentive mechanism with a selling price of 4c€. Note that this is the same as E1, but for the sake of clarity is used with the two names.
  - E7: environment with 100% volumetric charges. NB is used as an incentive mechanism with a selling price of 6c€.
  - E8: environment with 100% volumetric charges. NB is used as an incentive mechanism with a selling price of 8c€.
  - E9: environment with 100% volumetric charges. NB is used as an incentive mechanism with a selling price of 10c€.

For all of the environments, we set the value of $\Pi^{(en)}$ to 0.132 €/kWh. Furthermore, we assume an initial distribution tariff $\Pi_n^{(dis)}$ of 0.088 €/kWh (making an equivalent retail price of 0.22 €/kWh). To determine two-part tariffs (E2 - E4), we compute:

$$\forall i \in \mathcal{I} \quad c_i = \frac{(\Omega + \Xi_n) \cdot \left(\Pi_n^{(dis)} \cdot \eta\right)}{I + J} \cdot \gamma_i, \tag{8}$$

where η is the percentage of volumetric charges, and $\gamma_{i,n}$ is an adjustment factor applied depending on the peak demand of the consumer/prosumer, which is useful to charge users fixed costs depending on their power consumption. In this case study, γ_i is assumed equal to 1 for all prosumers.

5.1. Results

To represent the distribution network evolution for each environment, we rely on four metrics: (i) the evolution of the variable (volumetric) term of the distribution tariff, (ii) the penetration of DER relative to the maximum potential I, (iii) the actual deployed PV and battery capacity (in kWp and kWh), and (iv) the LCOE of the deployed DER installations (in €/kWh).

5.1.1. Tariff Designs (E1–E4)

According to Figure 3, the upper-left subfigure, the variable (volumetric) term of the distribution tariff increases in a quicker fashion when the share of this term in the two-part tariff design is large. Likewise, the deployment of DER over time (Figure 3, lower-left subfigure) and the actual DER deployed capacity (Figure 4, upper subfigure), which represents the counterpart to the growth of the distribution tariff, increase in environments where the variable term in the two-part design is more prominent. In Figure 5, we can observe that the probability density functions of environments E1–E4 exhibit larger installation sizes for E1 (note that E1 and E6 represent the same environment) than E2, E3, and E4. Finally, regarding the LCOE of the DER installations, the four cases' costs are similar to the equivalent retail price. Note that higher volume shares (E1) results in lower LCOEs. Finally, in Figure 6, the resulting LCOE of the prosumers is showcased. The red, dotted line indicates the electricity price (at the initialisation conditions) in €/kWh, without DER installation (i.e., for consumers).

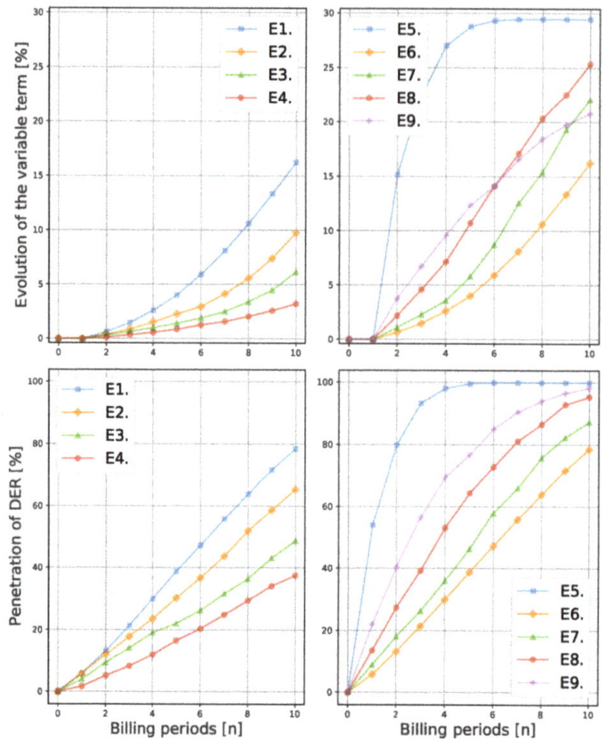

Figure 3. Evolution of $\Pi_n^{(dis)}$ (upper two figures) and of the DER adoption (lower two figures) across the discrete time dynamical system, for the evaluation of tariff designs E1–E4 (left-hand side figures), and of the incentive mechanisms E5–E9 (right-hand side figures).

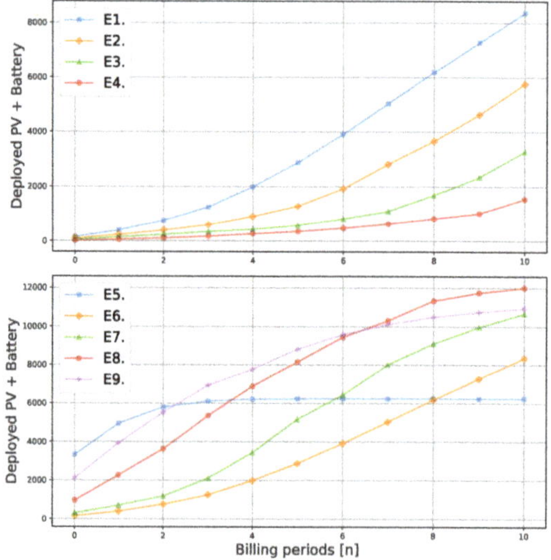

Figure 4. Cumulative sum of the size of the deployed DER installations (including PV and batteries), over the discrete time dynamical system. The upper figure corresponds to the evaluation of tariff designs, whereas the lower one corresponds to the evaluation of the incentive mechanisms.

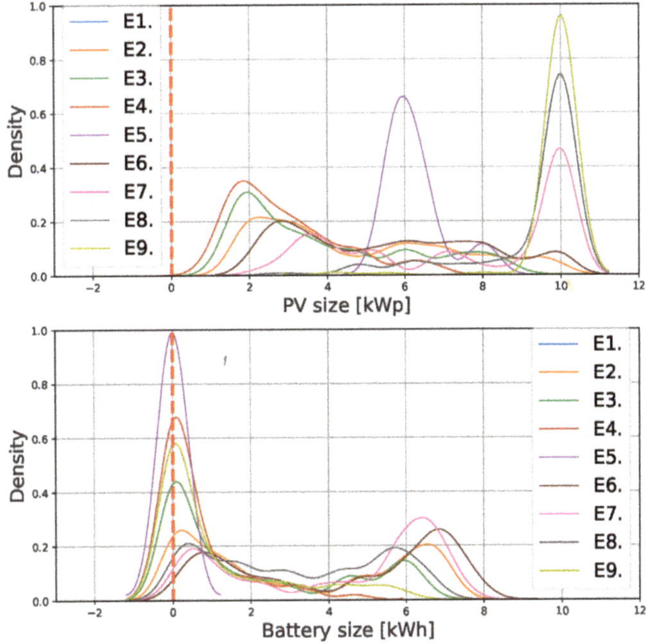

Figure 5. Gaussian kernel density estimation of the installed capacity of PV (upper plot), and of batteries (lower plot). These figures represent the probability density function for the kernel density estimation of PV and battery capacities, for every environment (E1–E9). This probability is computed based on the calculated DER installation size of the set $\mathcal{I}$.

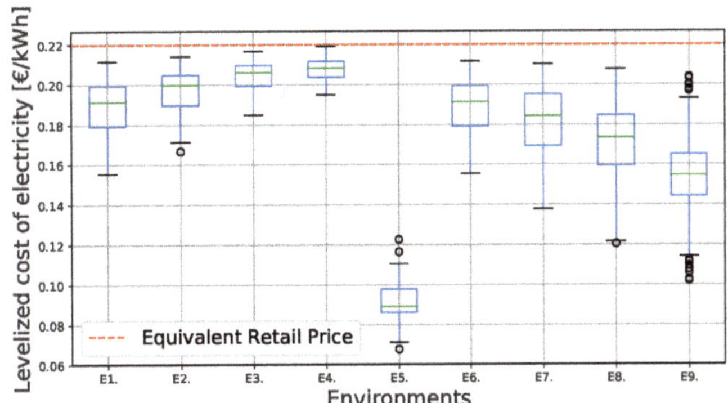

Figure 6. Levelized cost of electricity of the prosumers in set $\mathcal{I}$, for every environment (E1–E9).

5.1.2. Incentive Mechanisms (E5–E9)

Figure 3, upper-right subfigure, shows two different trends, one for the NM environment (E5), and another for the rest. E5 variable term outgrows the other four by at least 5%, followed by E8, E7, E9, and E6 at the end ($n = 10$) of the simulation. The same trends are observed in Figure 3, lower-right subfigure, which represents the total DER penetration. However, examining the total capacities of deployed DER (Figure 4, lower subfigure, and Figure 5), it is visible that, despite the larger DER penetration, E5 results in lower total capacity of deployed PV and batteries. Regarding the LCOE, E5 displays a considerably lower LCOE than the rest of the environments. Figure 6 displays the resulting LCOE of the prosumers for scenarios E5 to E9, as well as for the previous ones (see Section 5.1.1 for more details).

5.2. Discussion

5.2.1. Tariff Designs (E1–E4)

We observe unity in the results, by increasing the share of the variable term in the distribution tariff, the business case to deploy DER installations improves, thus facilitating the transition consumer → prosumer. This, in turn, causes the distribution tariff to grow further in the environments with higher share of variable term, indicating a larger potential death spiral behaviour for those environments. Hence, introducing a two-part design reduces the instability of the system, as already highlighted in [31]. If we observe the total amount of PV and battery deployed (Figure 4), we can deduce that relying on distribution tariffs which are predominantly volumetric results in larger deployed DER capacities. This suggests a trade-off between DER penetration and total capacity installed, and a distribution price spiral. Such a trade-off must be addressed by policy makers in order to decide the desired trend. Finally, since the incentive mechanism in place (NB with a selling price of 0.04 €/kWh) does not significantly improve the DER business case, the four LCOEs are similar to the equivalent retail price, as can be seen in Figure 6. The lowest LCOE corresponds to E1, which is consistent with Figures 4 and 5.

5.2.2. Incentive Mechanisms (E5–E9)

The different trends observed for NM and NB are a consequence of the distinct behaviour of prosumers they induce. With NM there is no incentive to make a business case selling electricity or becoming self-sufficient. NM offers the perfect scenario for the prosumers to adjust their production so that they import and export equivalent amounts of energy ($\xi_{i,n} = 0$). For this reason, the variable term in E5 (Figure 3), outgrows the other four environments, since the apparent consumption with E5 is close to zero, and the DSO needs to adjust the distribution tariff in a larger extent. We may also note that, under NM, no batteries are deployed (Figure 5). This is compatible with the findings in [5], where the authors observe that, with this system, batteries and imports are perfect substitutes. In Figure 6, we can the LCOE of these environments. The low LCOE of E5 is also consequence of the extremely low apparent consumption of the prosumers under NM. In the other four environments, the prosumers tend to deploy more PV and battery capacity to reduce their imports. Interestingly, when the selling price is high (E9), the prosumers rely on selling electricity as a business case, not reducing their apparent consumption in the same extent as E7 or E8. Hence, the increase in the distribution tariff is not so prominent in E9. A new trade-off appears between selling price and a distribution price spiral, where both imply an extra burden for the community.

6. Conclusions

In the context of increasing decentralised electricity generation, this paper has evaluated the effects of different regulatory frameworks on the evolution of distribution networks. A multi-agent model is used to simulate the behaviour of the agents of a distribution network. The actions of the agents are evaluated at several time-steps, leading to the evolution of the distribution network. Electricity consumers interacting with a single distribution network are modelled as rational agents that may invest in optimally sized distributed energy installations composed of PV and/or batteries. Finally, the distribution tariff is adapted according to the remuneration mechanism of the DSO.

We have designed and simulated four different examples based on gradually decreasing the proportion of volumetric charges switching them by fixed ones. Moreover, five different examples based on the type of incentive mechanism employed to promote DER are proposed. The results have been presented according to four distinct metrics: (i) the evolution of the volumetric term of the distribution tariff, (ii) the penetration of DER installations, (iii) the amount of deployed PV and batteries, and (iv) the LCOE of the deployed DER installations.

The results show that replacing volumetric charges with fixed ones impairs the economic business case of the consumers willing to deploy DER in the system. Regarding the use of different incentive mechanisms, using net-metering creates a potential spiral of the distribution tariff, with no integration of battery capacity in the system. When net-billing is used instead, the spiral of prices may be more easily contained by controlling the electricity selling prices. We observe a trade-off between spiralling electricity prices and the desired penetration of PV and batteries. Such trade-off may be tuned by policy makers by adjusting key parameters such as the level of capacity charges, or the selling price of electricity when net-billing is utilised.

This paper has analysed the different ways to foster DER adoption, and the regulatory challenges induced by them. Our approach is limited by the number of distribution network users who can be simulated, as well as by the assumption that only consumers whose electricity costs will decrease when deploying a DER installation may become prosumers. Future works could address a situation where the DSO allowed revenues do not remain constant throughout the simulation horizon; moreover, the use of non-rational agents may be addressed in the future.

Author Contributions: Conceptualization, D.E. and A.G.; methodology, M.M.d.V. and R.F.; software, M.M.d.V.; validation, M.M.d.V. and R.F.; investigation, M.M.d.V. and R.F.; resources, D.E.; data curation, M.M.d.V. and R.F.; writing—original draft preparation, M.M.d.V.; writing—review and editing, M.M.d.V., R.F., A.G. and D.E.; supervision, A.G. and D.E.; project administration, A.G.; funding acquisition, A.G. and D.E.

Funding: This research was funded by the public service of Wallonia within the framework of the TECR project.

Conflicts of Interest: The authors declare no conflict of interest. The funders had no role in the design of the study; in the collection, analyses, or interpretation of data; in the writing of the manuscript, or in the decision to publish the results.

Abbreviations

The following abbreviations are used in this manuscript:

DER	Distributed energy resources
PV	Photovoltaic
FiT	Feed-in tariff
FiP	Feed-in premium
DSO	Distribution system operator
DG	Distributed generation
NM	Net-metering
NB	Net-billing
MILP	Mixed integer linear program
LCOE	Levelized cost of electricity
E#	Environment number

References

1. *Study on the Effective Integration of Demand Energy Recourses for Providing Flexibility to the Electricity System;* Final report to The European Commission; European Commission: Brussels, Belgium, 2015; p. 179.
2. European Parliament, Council of the European Union. Directive 2009/28/EC of the European Parliament and of the Council of 23 April 2009 on the promotion of the use of energy from renewable sources and amending and subsequently repealing Directives 2001/77/EC and 2003/30/EC. *Off. J. Eur. Union* **2009**, *5*, 39–85.
3. Candelise, C.; Winskel, M.; Gross, R.J. The dynamics of solar PV costs and prices as a challenge for technology forecasting. *Renew. Sustain. Energy Rev.* **2013**, *26*, 96–107. [CrossRef]
4. Lopes, J.P.; Hatziargyriou, N.; Mutale, J.; Djapic, P.; Jenkins, N. Integrating distributed generation into electric power systems: A review of drivers, challenges and opportunities. *Electr. Power Syst. Res.* **2007**, *77*, 1189–1203. [CrossRef]

5. Gautier, A.; Jacqmin, J.; Poudou, J.C. The prosumers and the grid. *J. Regul. Econ.* **2018**, *53*, 100–126. [CrossRef]
6. Eid, C.; Guillén, J.R.; Marín, P.F.; Hakvoort, R. The economic effect of electricity net-metering with solar PV: Consequences for network cost recovery, cross subsidies and policy objectives. *Energy Policy* **2014**, *75*, 244–254. [CrossRef]
7. Jargstorf, J.; Belmans, R. Multi-objective low voltage grid tariff setting. *IET Gener. Transm. Distrib.* **2015**, *9*, 2328–2336. [CrossRef]
8. Costello, K.W.; Hemphill, R.C. Electric utilities''death spiral': Hyperbole or reality? *Electr. J.* **2014**, *27*, 7–26. [CrossRef]
9. Román, J.; Gómez, T.; Muñoz, A.; Peco, J. Regulation of distribution network business. *IEEE Trans. Power Deliv.* **1999**, *14*, 662–669. [CrossRef]
10. Comnes, G.A. *Review of Performance-Based Ratemaking Plans for US Gas Distribution Companies;* Electricity Markets and Policy Group: Berkeley, CA, USA, 1994.
11. Barker, P.P.; De Mello, R.W. Determining the impact of distributed generation on power systems. I. Radial distribution systems. In Proceedings of the Power Engineering Society Summer Meeting, Seattle, WA, USA, 16–20 July 2000; Volume 3, pp. 1645–1656.
12. Ackermann, T.; Andersson, G.; Soder, L. Electricity market regulations and their impact on distributed generation. In Proceedings of the International Conference on Electric Utility Deregulation and Restructuring and Power Technologies. Proceedings (Cat. No.00EX382), London, UK, 4–7 April 2000; pp. 608–613.
13. Pepermans, G.; Driesen, J.; Haeseldonckx, D.; Belmans, R.; D'haeseleer, W. Distributed generation: Definition, benefits and issues. *Energy Policy* **2005**, *33*, 787–798. [CrossRef]
14. Reneses, J.; Ortega, M.P.R. Distribution pricing: Theoretical principles and practical approaches. *IET Gener. Transm. Distrib.* **2014**, *8*, 1645–1655. [CrossRef]
15. Council of European Energy Regulators. *Electricity Distribution Network TariffsCEER Guidelines of Good Practice;* Technical Report; CEER: Brussels, Belgium, 2017.
16. Ortega, M.P.R.; Pérez-Arriaga, J.I.; Abbad, J.R.; González, J.P. Distribution network tariffs: A closed question? *Energy Policy* **2008**, *36*, 1712–1725. [CrossRef]
17. Pérez-Arriaga, I.J.; Jenkins, J.D.; Batlle, C. A regulatory framework for an evolving electricity sector: Highlights of the MIT utility of the future study. *Econ. Energy Environ. Policy* **2017**, *6*, 71–92.
18. Cossent, R.; Gómez, T.; Frías, P. Towards a future with large penetration of distributed generation: Is the current regulation of electricity distribution ready? Regulatory recommendations under a European perspective. *Energy Policy* **2009**, *37*, 1145–1155. [CrossRef]
19. European Commission. *Study on Tariff Design for Distribution Systems;* Technical Report; European Commission: Brussels, Belgium, 2015.
20. De Joode, J.; Jansen, J.; Van der Welle, A.; Scheepers, M. Increasing penetration of renewable and distributed electricity generation and the need for different network regulation. *Energy Policy* **2009**, *37*, 2907–2915. [CrossRef]
21. Frías, P.; Gómez, T.; Cossent, R.; Rivier, J. Improvements in current European network regulation to facilitate the integration of distributed generation. *Int. J. Electr. Power Energy Syst.* **2009**, *31*, 445–451. [CrossRef]
22. Faerber, L.A.; Balta-Ozkan, N.; Connor, P.M. Innovative network pricing to support the transition to a smart grid in a low-carbon economy. *Energy Policy* **2018**, *116*, 210–219. [CrossRef]
23. Picciariello, A.; Reneses, J.; Frias, P.; Söder, L. Distributed generation and distribution pricing: Why do we need new tariff design methodologies? *Electr. Power Syst. Res.* **2015**, *119*, 370–376. [CrossRef]
24. Sotkiewicz, P.M.; Vignolo, J.M. Towards a cost causation-based tariff for distribution networks with DG. *IEEE Trans. Power Syst.* **2007**, *22*, 1051–1060. [CrossRef]
25. González, A.; Gomez, T. Use of system tariffs for distributed generators. In Proceedings of the 16th Power Systems Computation Conference, Glasgow, UK, 14–18 July 2008.
26. Christoforidis, G.; Panapakidis, I.; Papadopoulos, T.; Papagiannis, G.; Koumparou, I.; Hadjipanayi, M.; Georghiou, G. A model for the assessment of different net-metering policies. *Energies* **2016**, *9*, 262. [CrossRef]
27. Laws, N.D.; Epps, B.P.; Peterson, S.O.; Laser, M.S.; Wanjiru, G.K. On the utility death spiral and the impact of utility rate structures on the adoption of residential solar photovoltaics and energy storage. *Appl. Energy* **2017**, *185*, 627–641. [CrossRef]

28. Darghouth, N.R.; Barbose, G.; Wiser, R.H. Customer-economics of residential photovoltaic systems (Part 1): The impact of high renewable energy penetrations on electricity bill savings with net metering. *Energy Policy* **2014**, *67*, 290–300. [CrossRef]
29. Darghouth, N.R.; Wiser, R.H.; Barbose, G.; Mills, A.D. Net metering and market feedback loops: Exploring the impact of retail rate design on distributed PV deployment. *Appl. Energy* **2016**, *162*, 713–722. [CrossRef]
30. Gautier, A.; Jacqmin, J. PV adoption in Wallonia: The role of distribution tariffs under net metering. In Proceedings of the 7th Conference on the Regulation of Infrastructures, Bordeaux, France, 21–22 June 2018.
31. Simshauser, P. Distribution network prices and solar PV: Resolving rate instability and wealth transfers through demand tariffs. *Energy Econ.* **2016**, *54*, 108–122. [CrossRef]
32. Lesser, J.A.; Su, X. Design of an economically efficient feed-in tariff structure for renewable energy development. *Energy Policy* **2008**, *36*, 981–990. [CrossRef]
33. Pyrgou, A.; Kylili, A.; Fokaides, P.A. The future of the Feed-in Tariff (FiT) scheme in Europe: The case of photovoltaics. *Energy Policy* **2016**, *95*, 94–102. [CrossRef]
34. Sioshansi, R. Retail electricity tariff and mechanism design to incentivize distributed renewable generation. *Energy Policy* **2016**, *95*, 498–508. [CrossRef]
35. Yamamoto, Y. Pricing electricity from residential photovoltaic systems: A comparison of feed-in tariffs, net metering, and net purchase and sale. *Sol. Energy* **2012**, *86*, 2678–2685. [CrossRef]
36. Fridgen, G.; Kahlen, M.; Ketter, W.; Rieger, A.; Thimmel, M. One rate does not fit all: An empirical analysis of electricity tariffs for residential microgrids. *Appl. Energy* **2018**, *210*, 800–814. [CrossRef]
37. Shang, Y. Hybrid consensus for averager–copier–voter networks with non-rational agents. *Chaos Solitons Fractals* **2018**, *110*, 244–251. [CrossRef]
38. Abdelmotteleb, I.; Gómez, T.; Ávila, J.P.C.; Reneses, J. Designing efficient distribution network charges in the context of active customers. *Appl. Energy* **2018**, *210*, 815–826. [CrossRef]
39. Manuel de Villena, M.; Gautier, A.; Fonteneau, R.; Ernst, D. A multi-agent approach to model the interaction between distributed generation deployment and the grid. In Proceedings of the CIRED Workshop 2018, Ljubljana, Slovenia, 7–8 June 2018.
40. Prata, R.; Carvalho, P.M. Self-supply and regulated tariffs: Dynamic equilibria between photovoltaic market evolution and rate structures to ensure network sustainability. *Util. Policy* **2018**, *50*, 111–123. [CrossRef]
41. Schittekatte, T.; Momber, I.; Meeus, L. Future-proof tariff design: Recovering sunk grid costs in a world where consumers are pushing back. *Energy Econ.* **2018**, *70*, 484–498. [CrossRef]
42. Manuel de Villena, M.; Fonteneau, R.; Gautier, A.; Ernst, D. Exploring Regulation Policies in Distribution Networks through a Multi-Agent Simulator. In Proceedings of the YRS2018, Brussels, Belgium, 24–25 May 2018.

Article

Control Strategy for a Grid Connected Converter in Active Unbalanced Distribution Systems

Boris Dumnic [1,*], Bane Popadic [1], Dragan Milicevic [1], Nikola Vukajlovic [1] and Marko Delimar [2]

[1] University of Novi Sad, Faculty of Technical Sciences, 21000 Novi Sad, Serbia; banep@uns.ac.rs (B.P.); milicevd@uns.ac.rs (D.M.); nikolavuk@uns.ac.rs (N.V.)
[2] University of Zagreb, Faculty of Electrical Engineering and Computing, 10000 Zagreb, Croatia; marko.delimar@fer.hr
* Correspondence: dumnic@uns.ac.rs; Tel.: +381-21-485-2558

Received: 13 March 2019; Accepted: 3 April 2019; Published: 9 April 2019

Abstract: The development in distributed energy resources technology has led to a significant amount of non-linear power electronics converters to be integrated in the power system. Although this leads to a more sustainable system, it also can have adverse impacts on system stability and energy power quality. More importantly, the majority of the distribution power systems currently are unbalanced (with asymmetrical voltages) due to load unbalance, while the most common fault types are unbalanced grid faults that can have many adverse effects on distributed resource operations. In that regard, proper control of the grid connected converters in active unbalanced distribution systems will become very important. This paper aims to present the behavior of the advanced grid connected converter control technique under different voltage states at the point of common coupling (according to the ABC classification). The main insufficiencies of the classical control technique will be highlighted, while the paper will propose an appropriate solution for mitigation of negative sequence currents under asymmetrical voltages at the point of common coupling. An extensive experimental verification of the proposed techniques is performed using an advanced laboratory prototype for research in grid integration of distributed resources. The experimental verification clearly demonstrates the benefits offered by the advanced control strategy.

Keywords: Active distribution systems; voltage asymmetry; converter control

---

1. Introduction

Recent advances in power electronic converter technologies and their subsequent involvement in modern renewable energy sources (RES) offered the possibility for the proliferation of power electronics based distributed energy resources (DERs) in the network [1]. Driven by strong economic and technical incentives, a trend of an increase in the installed capacity can be noted on a regional and a worldwide level, as seen in Figure 1 (for the European region).

However, with their intermittent nature, DERs can significantly influence the system stability and reliability, especially at the distribution voltage levels. Therefore, in order to motivate the operator to shift towards a more decentralized active distribution system (ADS) and move closer to more sustainable smart grid concepts [2], the grid connected converter (GCC) will have to assume the responsibility for fulfilling the grid interconnection requirements, i.e., the grid code (GC). As the key component of future power generation systems, the GCC will inevitably be responsible for tasks regarding energy trading, offering fault ride through (FRT) capabilities, and achieving the required power quality for supplying consumers.

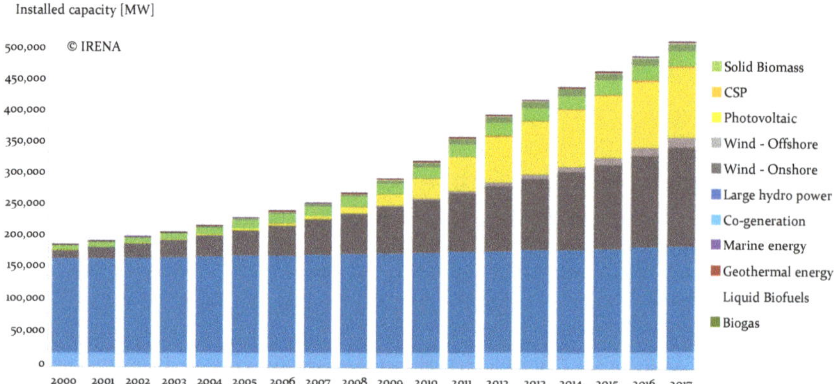

Figure 1. The cumulative installed capacity of renewable energy sources in Europe [3].

On the other hand, achieving such a predefined set of regimes (relevant to the respective GC) under unrated grid voltage conditions can prove to be a difficult task for classical GCC control techniques, especially during unbalanced grid faults. Moreover, many upcoming distribution networks can frequently have voltage asymmetry in the "normal" operating regimes due to unbalanced loads [4–6]. This may present the standard GCC control technique with an issue that they cannot resolve, since they are usually not well equipped to deal with asymmetrical grid voltages at the point of common coupling (PCC).

The control strategy of the GCC, as presented in [7], based on the instantaneous power theory (IPT), can take many different routes (by different proposed strategies), but essentially takes the two following directions—power characteristic oriented control or voltage support oriented control. It should be emphasized that within the different control strategies presented in [8–16], special attention has always been reserved for balanced current control. In most of the literature, as in [8], elimination of the negative sequence current is explained (within the IPT), however, the important issue of cross-coupling between positive and negative sequences has not been discussed. Similarly, in [10] and [11], the stationary reference frame (with the well-known issue of initial conditions) was used for the control of currents, consequently leaving out the discussion on the decoupling between positive and negative sequences. Similarly, an linear-quadratic regulator (LQR) controller with resonant elements in state feedback control was used in [12], while a virtual admittance controller as part of a synchronous power controller was proposed in [13]. The testing of a relatively complex disturbance observer with a second order general integrator (SOGI) was performed in [14] using only simulation. The authors in [15] used generalized sinusoidal pulse-width modulation (SPWM) and a voltage source converter sequence subsystem to balance the current or mitigate DC-bus voltage ripples, avoiding classical controllers and resulting in the dynamic and steady state response being susceptible to parameter variation. In [16], the double decoupled synchronous reference frame phase-locked loop (PLL) was used for synchronization, with the negative sequence currents regulated using an additional notch filtering in both sequence current control loops. In [17–20], mostly harmonic distortion mitigation and synchronization under distorted and asymmetrical voltages were discussed. Considering the previous literature review, the general conclusion can be made that GCC control techniques usually lack the simplicity for implementation in real time that would allow new and existing DERs to operate indefinitely or according to the GC in active unbalanced distribution systems characterized by asymmetrical voltages at the PCC.

This paper aims to showcase the behavior of the standard GCC control technique in active unbalanced distribution systems. Presenting the influence of different grid voltage asymmetry parameters (based on a well-known classification of voltage sags) on the elements of GCC control, the focus will be set on power quality issues and the characteristics of currents injected at the PCC. Within this paper, the authors will explain the methods used for improvement of the main control subsystems that would allow the operation of the GCC under asymmetrical voltages. Subsystems, such as the GCC synchronization unit and GCC negative sequence current control, will be improved or introduced. The improved control technique uses delay signal cancelation (DSC) methodology, in order to differentiate between the positive and negative sequence of the voltage (for the synchronization) and the current (for negative sequence mitigation). Finally, the paper will present the results of an extensive experimental verification. The experiments will include testing the improved control strategy under a significant number of potential grid fault types and voltage asymmetry due to load unbalance.

The main research contributions presented in this paper are as follows:

- This paper presents an improved GCC control technique capable of maintaining adequate operation under asymmetrical voltages, showcasing the main benefits in comparison to the classical technique.
- The paper gives an extensive experimental verification of the improved technique behavior under different voltage sag types (according to the ABC classification), including operation under the most severe conditions. The developed GCC control technique is an improvement of the classical control technique with relatively low computational complexity and, in that regard, can be implemented in new GCC, as well as in the GCC already existing in the ADS.

2. Grid Connection Requirements in Active Unbalanced Distribution Networks

Systems that allow power flow management in a flexible network topology are, per the International Council on Large Electric Systems (CIGRE) C6.19 workgroup, defined as active distribution systems—an extension from active distribution networks [21]. The flexible topology refers to a combination of DERs, designated as generators in parallel operation with the distribution system, loads, or storage units [22]. Consequently, the power flow along the network feeder becomes arbitrary and has to be controlled by a centralized intelligent system that monitors the current network conditions. In such instances, the conventional power flow models become inapplicable [1,23]. Applications where the GCC are used in parallel operation with the ADS are shown in Figure 2. State variables in such cases can originate from the ADS operator or primary energy supply, and influence the control variables with different requirements.

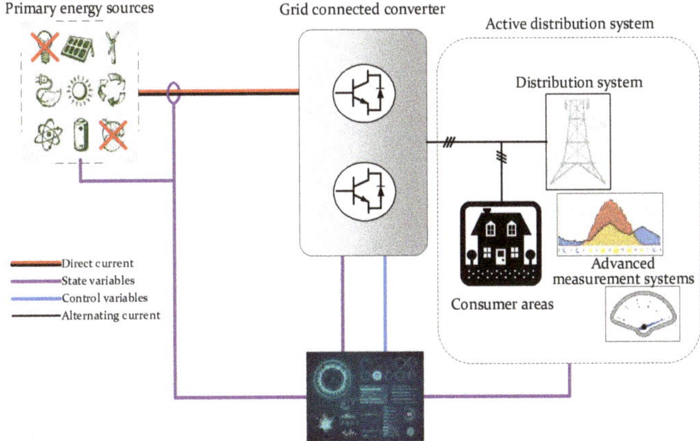

Figure 2. Grid connected converter applications in the active distribution system.

Undoubtedly, if high penetration of DERs is achieved with non-dispatchable generation units, ADSs would face more technical uncertainties than the classical distribution network [24]. Therefore, advanced RES systems will have to, as much as possible, abide by the traditional constraints regarding the connection requirements (power limitation, power quality, short-circuit current limitation). Under special circumstances, where it was impossible to uphold the traditional constraints, a special set of requirements was developed. In order to allow the distribution network operator (DNO) the possibility of ensuring adequate and reliable energy supply, these requirements are collected and published by the relevant DNO in a technical document known as the grid code. Furthermore, these requirements also provide the least possible duration of the connection process and the assurance for DER technology manufacturers during the design process. Usually, all currently available GCs require DERs to possess reactive energy injection, inertia, and oscillation damping to support system transience and steady state stability. Moreover, DERs need to offer voltage support under different grid faults, i.e., offer FRT capabilities. Dictated by the state of the own power system, these requirements may differ worldwide, but they represent the corner stone of every grid code [25]. Figure 3 represents the FRT requirements given by the relevant GCs in different countries.

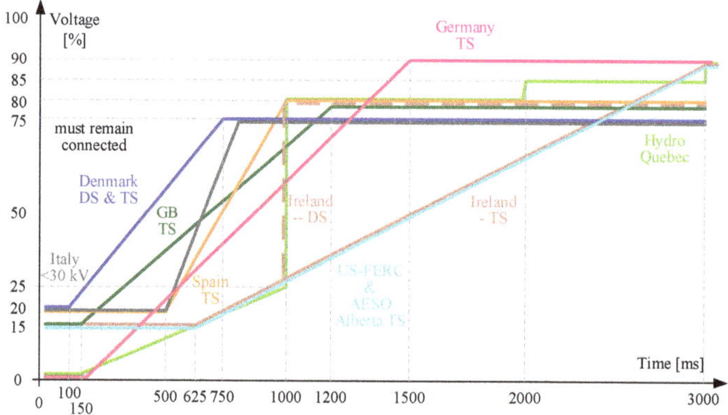

Figure 3. Fault ride-through requirements by different distribution systems' (DSs) and transmission systems' (TSs) grid codes.

In accordance with the given characteristics, the control of the GCC needs to fulfill two main requirements during grid faults: 1) To remain connected to the grid in the event of grid faults for a requested amount of time, and 2) to support the voltage at the point of common coupling by injecting a requested amount of reactive energy.

As an example, in the case of one of the most developed energy markets, the GCC needs to remain operational in an event of a grid fault for the following duration (given in milliseconds in accordance with the German GC):

$$t = 150 + \frac{1500 - 150}{0.9} \cdot (1 - \Delta U), \tag{1}$$

where ΔU designates the relative value of the voltage drop at the PCC. In the same grid code, the reactive current to be injected to the grid is calculated by the following equation:

$$i_q = \min(2 \cdot \Delta U \cdot i_r, i_{\max}). \tag{2}$$

Achieving these requirements is not a complex task for the classical control technique in case of symmetrical voltages and balanced grid faults at the PCC. However, when grid faults are unbalanced and voltages at the PCC become asymmetrical, the fulfillment of these requirements becomes a fairly hard task. Furthermore, the GCC is required to sustain operation indefinitely under asymmetrical voltages in case of unbalanced grid faults with less than 10% voltage drops or, more importantly, under asymmetrical voltages induced by unbalanced loading in the active distribution system. The most common and frequently used classification of grid voltage sags is the ABC classification based on a simplified network model [26,27]. For this particular classification, there are seven distinct types (A–G), as shown in Table 1, with each one having a different magnitude or phase angle variation. The rated voltage value at the PCC is denoted by V, while the factor of the voltage value contributing to the fault is denoted by k (relative voltage drop—can assume values between 0 and 1). While the most natural classification would assume the use of symmetrical components, it is rarely used due to comprehension complexity. It is clear that, albeit being based on incomplete assumptions (simplified network model), the ABC classification, F, is simpler and more convenient for use in different applications.

Table 1. The ABC classification of voltage sags.

Type	Voltage Value	Voltage Vector
A	$\overline{v}_a = (1-k)V\ \overline{v}_b = -\frac{1}{2}(1-k)V - j\frac{\sqrt{3}}{2}(1-k)V$ $\overline{v}_c = -\frac{1}{2}(1-k)V + j\frac{\sqrt{3}}{2}(1-k)V$	
B	$\overline{v}_a = (1-k)V\ \overline{v}_b = -\frac{1}{2}V - j\frac{\sqrt{3}}{2}V\ \overline{v}_c = -\frac{1}{2}V + j\frac{\sqrt{3}}{2}V$	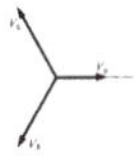
C	$\overline{v}_a = V\ \overline{v}_b = -\frac{1}{2}V - j\frac{\sqrt{3}}{2}(1-k)V\ \overline{v}_c = -\frac{1}{2}V + j\frac{\sqrt{3}}{2}(1-k)V$	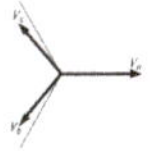
D	$\overline{v}_a = (1-k)V\ \overline{v}_b = -\frac{1}{2}(1-k)V - j\frac{\sqrt{3}}{2}V\ \overline{v}_c = -\frac{1}{2}(1-k)V + j\frac{\sqrt{3}}{2}V$	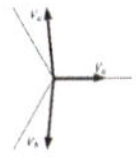
E	$\overline{v}_a = V\ \overline{v}_b = -\frac{1}{2}(1-k)V - j\frac{\sqrt{3}}{2}(1-k)V$ $\overline{v}_c = -\frac{1}{2}(1-k)V + j\frac{\sqrt{3}}{2}(1-k)V$	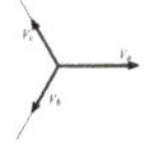
F	$\overline{v}_a = (1-k)V\ \overline{v}_b = -\frac{1}{2}(1-k)V - j(\frac{\sqrt{3}}{6}(1-k)V + \frac{\sqrt{3}}{3}V)$ $\overline{v}_c = -\frac{1}{2}(1-k)V + j(\frac{\sqrt{3}}{6}(1-k)V + \frac{\sqrt{3}}{3}V)$	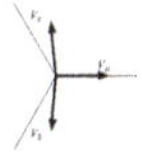
G	$\overline{v}_a = \frac{2}{3}V + \frac{1}{3}(1-k)V\ \overline{v}_b = -\frac{1}{3}V - \frac{1}{6}(1-k)V - j\frac{\sqrt{3}}{2}(1-k)V$ $\overline{v}_c = -\frac{1}{3}V - \frac{1}{6}(1-k)V + j\frac{\sqrt{3}}{2}(1-k)V$	

The voltage sag type, occurring at the PCC, will depend on different factors of the network topology and their combination. First and foremost, the type of the grid fault (balanced or unbalanced) will significantly influence the voltages at the fault location and therefore at the point of measurement (at the PCC). There are four distinct fault types that can occur in the distribution system: Single line to ground fault (SLG), line to line fault (LL), double line to ground fault (LLG), and three phase fault (3P). Balanced grid faults, causing symmetrical voltages, are exclusively related to 3P faults and classified as voltage sag type A. Depending on other factors, LL and SLG faults can result in voltage sag types of B, C, and D, while E, F, and G voltage sag types originate from LLG faults. The transformers, responsible for inner voltage transformation, can influence the voltage sag type, as experienced from the PCC. Since the per unit voltage of the primary and the secondary are equal for the transformers of the $Y_n y_n$ type, these transformers do not influence the voltage sag types at the PCC. In case of the Dd, Dz, and Yy transformers, the zero sequence voltage will be removed. The most important change comes from the Dy, Yd, and Yz transformers, since they change the line and phase voltages, resulting in an altered voltage sag type in the process. In the same context, the load connection (Y or D) will also significantly affect the voltage sag type at the PCC.

As a logical extension of the ABC classification, the voltage space vector classification was proposed and used in [27]. The basis for the extension is found in the amplitude invariant Clarke transform, where voltages are represented in a stationary reference frame by the following components: α, β, and *zero*. The main parameters that define the voltage (ellipse) in the stationary reference frame are the large radius (r_{max}), the small radius (r_{min}), and the inclination (φ_{inc}). The values of the parameters can be calculated from the positive (superscript p) and negative (superscript n) sequence voltages as [28]:

$$r_{max} = |V^p| + |V^n|;\ r_{min} = ||V^p| - |V^n||\ and\ \varphi_{inc} = \frac{1}{2}(\theta_p + \theta_n). \quad (3)$$

In order to fully classify the arbitrary voltage sag in regard to the ABC classification, the voltage space vector shape index (SI) can be defined as shown in Equation (4):

$$SI = \frac{r_{min}}{r_{max}} = \frac{||V^P| - |V^n||}{|V^p| + |V^n|}. \quad (4)$$

The value of the SI represents the compliance of the voltage space vector, with the circular shape achieved by symmetrical voltages at the PCC, where $SI = 1$ represents a circle, $SI = 0$ is a two dimensional line, and the ellipse has values in the following range, $0 < SI < 1$. Every combination of the values for these four parameters can be classified in one of seven main types of voltage sags (A–G).

3. Control Methodology for a Grid Connected Converter under Asymmetrical Voltages

In accordance with the IPT, the operation of the GCC under asymmetrical voltages at the PCC is significantly more complex for the classical control strategy. First and foremost, the synchronization unit, based on a simple PLL in a synchronous reference frame, is unable to properly estimate the position (angle) of the grid voltage vector represented [29]. The main issue with the most used synchronization technique is the loop filter that consists of a simple PI type controller, which is inadequate for non-constant value signals at its input. Clearly, when the voltage is asymmetrical, after applying the amplitude invariant Park transform, the resulting value in the synchronous rotating reference frame is a superposition of a constant (positive sequence) component and variable (negative sequence) component at twice the grid fundamental frequency, as observed in Figure 4 and Equations (5) and (6). The same reasoning holds true for the current control loop as well, where simple PI controllers in the classical technique are insufficient for the control of the desired GCC reference values [30].

$$\begin{bmatrix} v_d^p \\ v_q^p \end{bmatrix} = V^p \begin{bmatrix} cos(\theta^p) \\ sin(\theta^p) \end{bmatrix} + V^n cos(\theta^n) \begin{bmatrix} cos(2\omega t) \\ -sin(2\omega t) \end{bmatrix} + V^n sin(\theta^n) \begin{bmatrix} sin(2\omega t) \\ cos(2\omega t) \end{bmatrix}. \quad (5)$$

$$\begin{bmatrix} i_d^p \\ i_q^p \end{bmatrix} = I^p \begin{bmatrix} cos(\psi^p) \\ sin(\psi^p) \end{bmatrix} + I^n cos(\psi^n) \begin{bmatrix} cos(2\omega t) \\ -sin(2\omega t) \end{bmatrix} + I^n sin(\psi^n) \begin{bmatrix} sin(2\omega t) \\ cos(2\omega t) \end{bmatrix}. \quad (6)$$

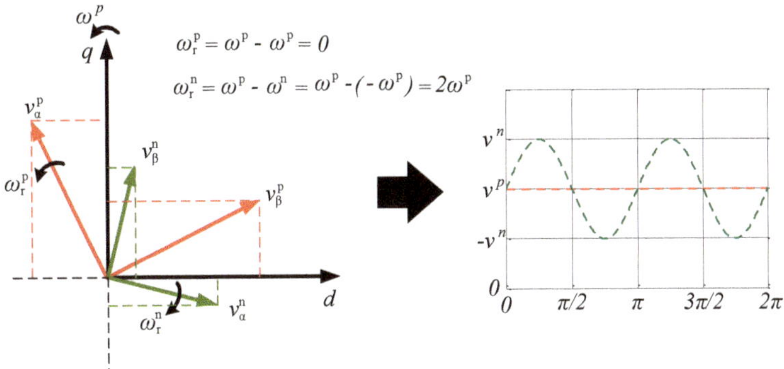

Figure 4. Asymmetrical value in a synchronous rotating reference frame (Park transform).

In general, when operating under asymmetrical voltages and currents, the power of the GCC can be given by Equation (7), presenting 4 degrees of freedom to control six variables (powers). However, in order to achieve proper power control of the GCC, decoupling between positive and negative sequences needs to be achieved. The delay signal cancelation technique can be used to separate the voltage and current sequences [31,32]. This offers the possibility of individual control of either the positive or negative sequence current, provided the correct reference frame alignment (estimation of the grid voltage vector phase angle). In the digital implementation, the DSC technique can be implemented according to Equations (8) and (9) for the positive and negative sequence, respectively (the expression holds true for both voltages and currents).

$$\underbrace{\begin{bmatrix} P \\ \widetilde{P}_1 \\ \widetilde{P}_2 \\ Q \\ \widetilde{Q}_1 \\ \widetilde{Q}_2 \end{bmatrix}}_{p-q} = \frac{3}{2} \underbrace{\begin{bmatrix} v_d^p & v_q^p & v_d^n & v_q^n \\ v_d^n & v_q^n & v_d^p & v_q^p \\ v_q^n & -v_d^n & -v_q^p & v_d^p \\ v_q^p & -v_d^p & v_q^n & v_d^n \\ v_q^n & -v_d^n & v_q^p & -v_d^p \\ -v_d^n & -v_q^n & v_q^p & v_d^p \end{bmatrix}}_{V_{dq}^{pn}} \begin{bmatrix} i_d^p \\ i_q^p \\ i_d^n \\ i_q^n \end{bmatrix}. \tag{7}$$

$$\underline{\hat{v}}_{dq}^p(t) = \frac{1}{2}\left[\underline{v}_{dq}(kT_s) + j\underline{v}_{dq}(kT_s - n_d T_S)\right], \tag{8}$$

$$\underline{\hat{v}}_{dq}^n(kT_S) = \frac{1}{2}\left[\underline{v}_{dq}(kT_S) - j\underline{v}_{dq}(kT_S - n_d T_S)\right], \tag{9}$$

where the parameter, n_d, equals a quarter of the ratio between the sampling and the grid frequencies. Due to known issues with the discretization of the DSC technique, special attention should be given to the selection of the sampling frequency, since the mentioned parameter, n_d, should be an integer number. If the discretization introduces a delay time of ΔT to the $n_d T_s$ product, the expected error for the negative sequence separation will be:

$$\Delta\underline{\hat{v}}_{dq}^n(kT_s) = \begin{bmatrix} V^p \cdot \sqrt{\frac{1}{2}(1-\cos(\frac{\pi}{2}\frac{\Delta T}{T_g/4}))} \cdot e^{j(2\omega\frac{\Delta T}{T_g/4}T_s - \frac{\pi}{2} + \theta^p - \arctan[\frac{\cos(\frac{\pi}{2}\frac{\Delta T}{T_g/4})-1}{\sin(\frac{\pi}{2}\frac{\Delta T}{T_g/4}}])} \\ +V^n \cdot \sqrt{\frac{1}{2}(1-\cos(\frac{\pi}{2}\frac{\Delta T}{T_g/4}))} \cdot e^{-j(\theta^n - \arctan[\frac{\cos(\frac{\pi}{2}\frac{\Delta T}{T_g/4})-1}{\sin(\frac{\pi}{2}\frac{\Delta T}{T_g/4})}] - \frac{\pi}{2})} \end{bmatrix}. \tag{10}$$

The first step in the proper control of GCC is the accurate estimation of the grid voltage vector phase angle in the direct sequence. However, when asymmetrical voltages are applied to the classical

PLL in the stationary reference frame, the loop filter is incapable of making the correct estimation of the angular frequency of the grid voltages. Therefore, an improvement of the classical PLL is necessary for GCC in active unbalanced distribution systems. The proposed DSC technique can be used in the phase detector stage of the PLL for the separation of positive sequence voltages. Additionally, the loop filter can be equipped with the resonant element in order to deal with the possible DSC discretization error and, more importantly, with the higher voltage harmonics at the PCC [29]. After the addition of the resonant element to the loop filter, the transfer function for the PLL in the synchronous reference frame becomes:

$$\Delta\omega_g = G_{fil}(s)\varepsilon(s) = \left(K_p + \frac{K_i}{s} + \frac{K_i s}{s^2 + \omega^2}\right)\varepsilon(s) \wedge \hat{\theta}_g = \frac{1}{s}\hat{\omega}_g(s), \tag{11}$$

while the overall outlook of the improved PLL unit is presented in Figure 5.

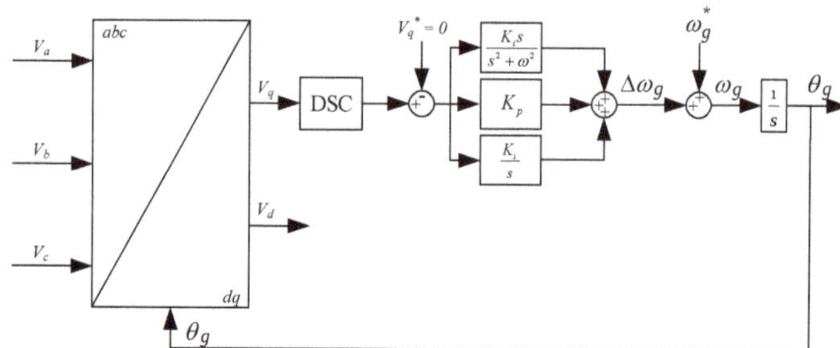

Figure 5. Improved phase-locked loop technique for asymmetrical voltages at the PCC.

The improved control technique should aim to achieve proper control of the GCC under asymmetrical grid voltages, with the least possible variation of the control strategy in regard to the classical strategy. To achieve such control, while remaining in full accordance to the GC requirements (for the injection of the direct sequence q-axis current), the most reasonable approach assumes mitigation of the negative sequence currents in the synchronous reference frame. If the mitigation of the negative sequence currents was possible, the remainder of the control technique (i.e., the positive sequence current and, therefore, the power control) can remain completely the same.

The improved GCC technique is presented in the Figure 6. All additional elements that were not part of the classical control strategy have a different background color (light blue). It is obvious that the DSC technique is added only to the negative sequence current controller and the PLL unit, leading to a negligible increase in processing resources. After mitigation of the negative sequence currents, the DSC technique in the positive sequence current control loop would be redundant. Therefore, the current controllers in both positive and negative sequences can be kept from the classical control, i.e., they can be PI type controllers. Furthermore, in the positive sequence current control loop, the parameters for the PI controllers can remain the same. For the PI controller parameters in the negative sequence, the parameters were recalculated using the modified symmetrical optimum method. Considering the fact that the negative sequence current is delayed by T_g, the integral gain parameter of the PI controller can be calculated as following:

$$K_i = \frac{1}{2\frac{1}{R_s}(T_i + \frac{T_g}{4})}, \tag{12}$$

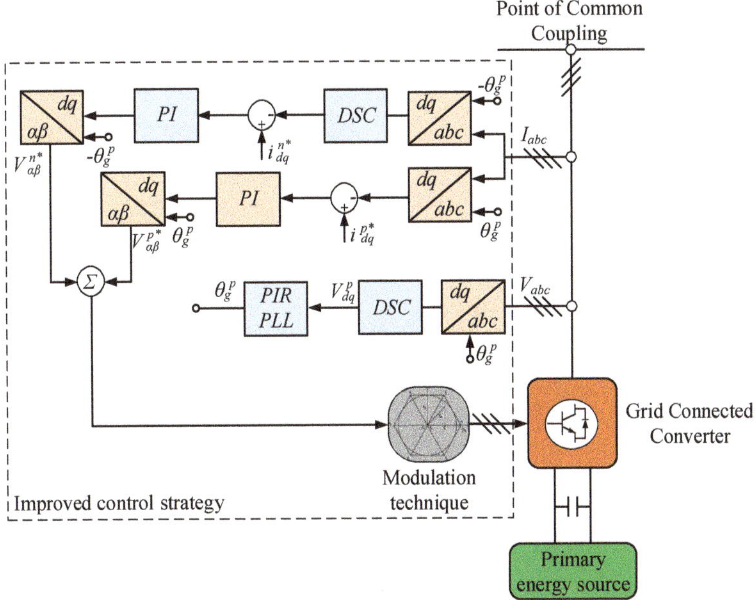

Figure 6. Improved GCC control technique with mitigation of the negative sequence currents.

The proportional gain of the controller is kept the same as for the positive sequence currents, i.e., it was calculated using the classical symmetrical optimum given by the following equation:

$$K_p = \frac{L_s}{2T_i}, \quad (13)$$

4. Experimental Verification of the Improved GCC Control Strategy

The behavior of the GCC control under asymmetrical grid voltages was tested on an advanced laboratory prototype, developed at the Faculty of Technical Sciences, University of Novi Sad [33]. The prototype represents a scaled model of the grid integrated DER and includes state-of-the-art hardware in the field of electrical drives and control, as shown in Figure 7.

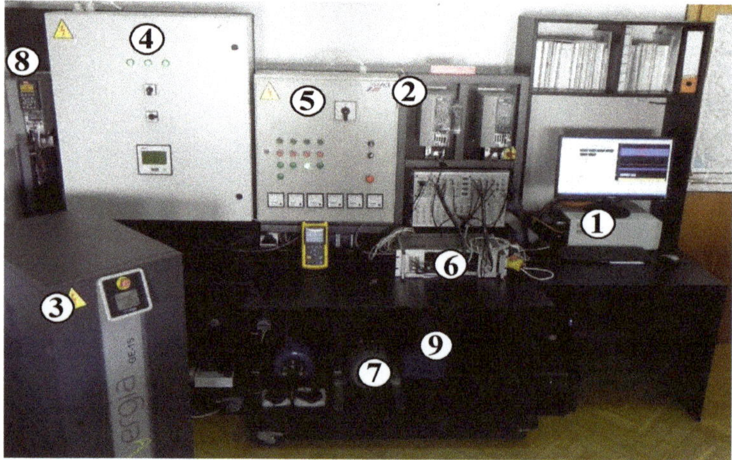

Figure 7. Laboratory prototype for testing grid integrated DERs.

It is based on a modular and highly versatile dSPACE processor board, marked with ①, where the control algorithm for the GCC is implemented. The GCC, marked with ②, is connected to the grid emulator ③, which provides adjustable grid voltages for experimental verification of the technique under different operating conditions (i.e., grid fault) at the PCC. Two distribution cabinets marked with ④ and ⑤ hold the switching and protection gear. Data acquisition, control signals, and measured signals are routed through the adapter block marked with ⑥. Furthermore, different grid integrated DER types (i.e., wind, marine, and solar energy conversion systems, co-generation systems, electrical energy-based storage systems, etc.) can be tested using various electrical machines ⑦, torque-controlled drives that emulate primary energy sources ⑧, and DC/DC ⑨ converters that complete the setup.

In order to compare the standard control technique and the improved control technique proposed within this paper, several experiments were performed, each aiming to show the behavior for different (asymmetrical) voltage conditions at the PCC.

4.1. Experiment I—10% Voltage Sag in One Phase

For the first experiment, the classical and improved technique was tested during a simple 10% voltage sag in one phase. The grid voltages waveforms at the PCC are presented in Figure 8, where obvious asymmetry in the voltages is present from the time of 5.1 s onward. This variation of voltage can be classified as a B type voltage sag, while for a magnitude of the voltage sag of only 10%, the GCC would be required to stay connected indefinitely. The current reference during the experiment was set to 3 A of the d-axis current. As expected, for the standard technique, the control was incapable of achieving a current reference when asymmetrical voltage conditions were applied at the PCC. As presented in Figure 9 (5.1 s), the currents were clearly asymmetrical, with two currents above (5 A) and one current below (1 A) the set-point value. This operation of the GCC would clearly be inadmissible and the GCC would need to disconnect, violating the GC in the process. Then, at the time of 7.7 s, the control of the GCC was switched from the classical to the improved strategy. Clearly, with a fast transient response, the improved control strategy was able to control the reference to the set-point value of 3 A in all phases. In that instance, full controllability of the GCC was achieved and the currents injected to the grid were symmetrical. The operation of the GCC under the improved technique could therefore be sustained indefinitely. Observing the voltage vector trajectory (Figure 10a), a slight deviation from the circular shape can be noted. However, for this slight voltage deviation, we can see that the current vector trajectory (Figure 10b) has a significant shape distortion during the voltage unbalance for the standard control technique, once again clearly supporting the previous statements. Figure 11a–c show the power, positive sequence currents, and negative sequence currents, respectively, during the first experiment. An oscillating component for both the power and positive sequence currents can be observed during operation under the classical control technique and asymmetrical voltages. The oscillating components correspond to the negative sequence currents as proposed by the theory. After the addition of the improved control strategy, the negative sequence currents are clearly mitigated, which was confirmed by the absence of oscillations in the positive sequence currents. This allows the control technique to remain completely the same for the positive sequence current control. On the other hand, severely reduced oscillations still exist in the power, since the voltages at the PCC are still asymmetrical.

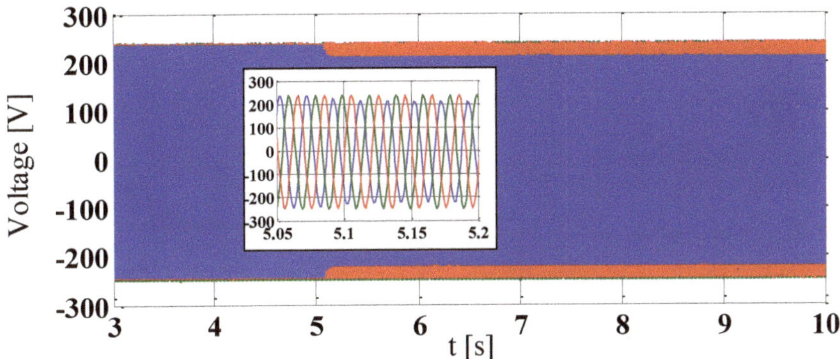

Figure 8. Grid voltage waveforms at the point of common coupling—10% voltage sag in one phase.

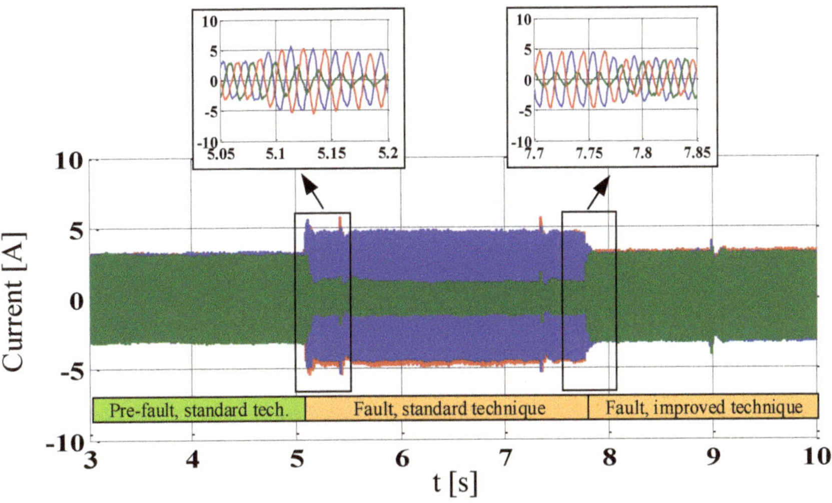

Figure 9. Converter (grid injected) current waveforms at the point of common coupling in one phase for the standard and improved technique—10% voltage sag in one phase.

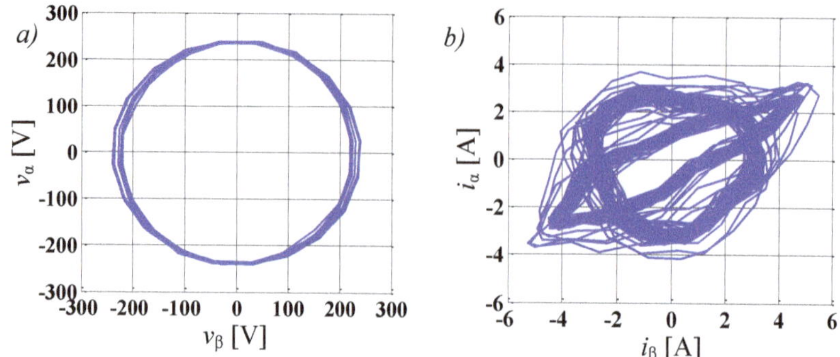

Figure 10. Voltage (**a**) and current (**b**) space vector—10% voltage sag in one phase.

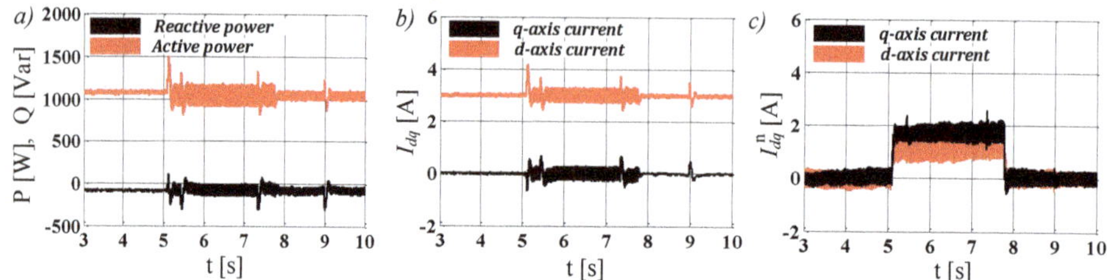

Figure 11. Grid connected converter power (**a**), positive sequence current (**b**), and negative sequence current (**c**) during the experiment—10% voltage sag in one phase.

4.2. Experiment II—20% Voltage Sag in One Phase and a 5 Degree Phase Shift

In the next experiment, the GCC standard control technique was tested for a one phase 20% voltage sag with a phase shift of 5 degrees, as seen in Figure 12a. The *d*-axis current reference was set to 2 A and at the time of 12.76 s, the described voltage sag occurred at the PCC. In Figure 12b–e, the GCC power, positive sequence currents, and negative sequence currents are presented respectively. It is evident that the standard control technique was incapable of maintaining the controllability of the system, since the injected currents are asymmetrical and significantly higher than the set-point value. The influence of the phase shift on the GCC operation has far more significant consequences, evident from the current waveforms in Figure 12b, where the current amplitude for the two phases was 11 A. The oscillatory component in the injected power and direct current components also increased. This is a consequence of such a high negative sequence component (Figure 12e), with 10 A in the negative *d*-axis and 2 A in the negative *q*-axis. Clearly, operation of the converter would be unacceptable, especially considering that the currents almost reached the protective devices' current limitation (14 A) of the tested GCC. To fully demonstrate the severity of the fault and the incapability of the standard technique, Figure 12f shows the operation of the respective GCC under the new experiment with 3 A of reference prior to the fault. When the same fault occurred at 12.87 s, the injected current shot over the set current protection limit (14 A) and the protective device switched off the GCC. In that regard, it would be impossible to operate according to the GC using the standard technique.

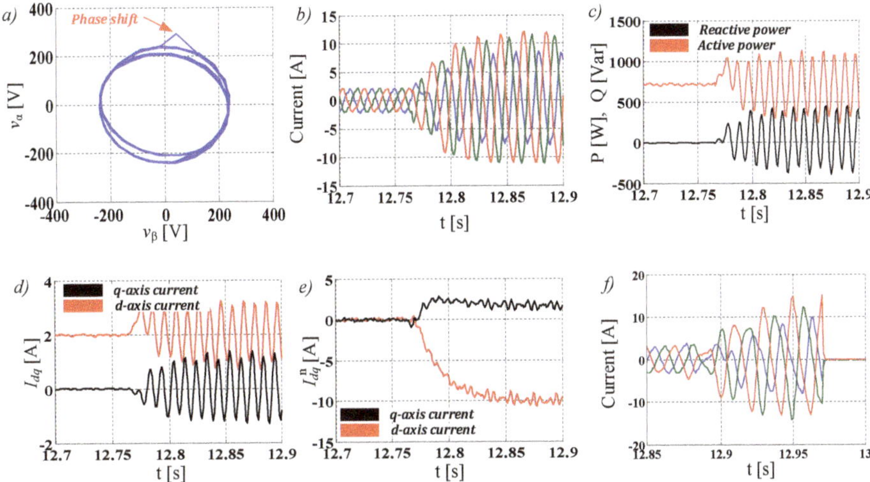

Figure 12. Grid connected converter operation under one phase 20% voltage sag with a phase shift of 5 degree—voltage space vector trajectory (**a**), current waveforms for 2 A reference (**b**), power for 2 A reference (**c**), positive sequence currents for 2 A reference (**d**), negative sequence current for 2 A reference (**e**), and current waveforms for 3 A reference (**f**).

On the other hand, the improved control technique had no issues operating under a 20% voltage sag with a 5 degree phase shift, even for the 3 A reference, as shown in Figure 13. The improved technique had no issues with the current protection limitation, while the currents were symmetrical both prior and after the grid fault. Since the negative sequence currents were completely mitigated and the oscillations in the injected power were significantly reduced, the operation of the GCC is fully controllable and could be sustained indefinitely or according to the requirements in the GC.

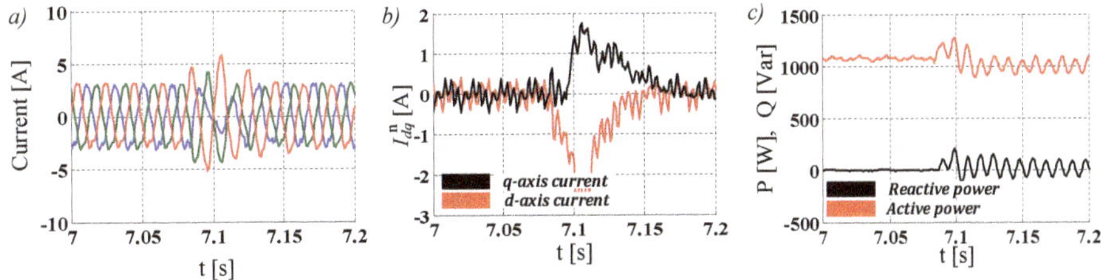

Figure 13. Improved control technique operation for a one phase 20% voltage sag with a 5 degree phase shift—current waveforms (**a**), negative sequence currents (**b**), and power (**c**).

4.3. Experiment III and IV—Voltage Sags Type C and E

Further verification of the improved control strategy was performed for voltage sags type C and type E presented in Figures 14 and 15, respectively. The selected type C voltage sag had a parameter of $k = 0.1$ (two-phase voltage sag of 8% with the 3 degrees phase shift), while for the type E voltage sag, $k = 0.4$ (two-phase voltage sag of 40%), according to Table 1. Observing Figures 14 and 15, it is easy to conclude that the improved technique can extend the range of operation for the GCC, even for the most severe of the voltage sag types. The injected GCC currents are symmetrical, and the negative sequence current is fully mitigated, while the power is controlled, with little oscillation consequent to the grid voltage asymmetry.

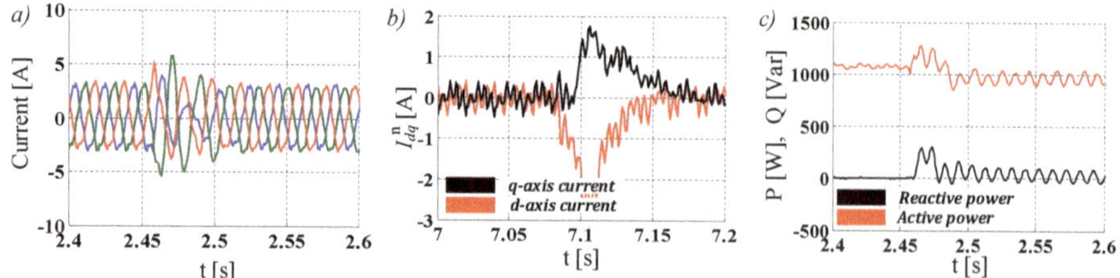

Figure 14. Improved control technique operation for the C type voltage sag—current waveforms (**a**), negative sequence currents (**b**), and power (**c**).

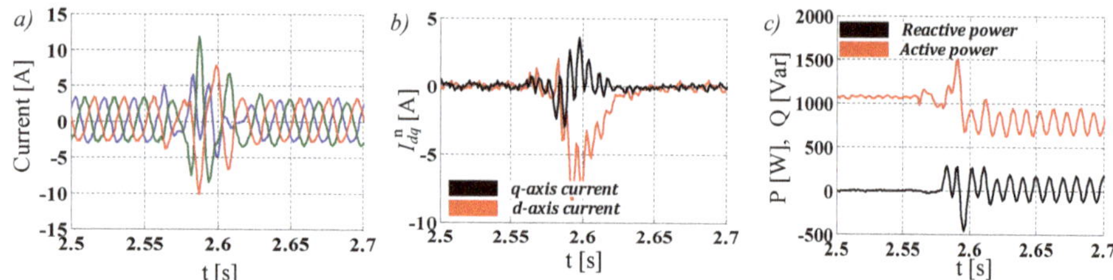

Figure 15. Improved control technique operation for the E type voltage sag—current waveforms (**a**), negative sequence currents (**b**), and power (**c**).

4.4. Experiment V—10% Overvoltage in One Phase

Finally, the techniques were tested against asymmetrical overvoltage at the PCC. The level of the voltage increase is directly related to the distribution network strength (or the network impedance) as observed at the PCC. If the distribution feeder strength is comparable with the integrated DERs power (i.e., high impedance of the network), the voltage at the PCC will consequently begin to increase. In that regard, Figure 16 presents the GCC phase currents for the classical and the improved technique during the 10% overvoltage in one phase, which would also require the GCC to remain connected indefinitely. The reference for the *d*-axis current was set at 3 A prior to the fault that occurred at the time of 4.75 s. The classical control technique had the expected behavior, with the asymmetrical currents of 5 A (two phases) and less than 1 A (one phase) injected to the grid at the PCC. After the application of the improved control technique, full controllability was restored, with symmetrical currents clearly regulated to the desired set-point. The shift in voltage trajectory, presented in Figure 17a, can be regarded as insignificant. However, after observing the current vector trajectory in the Figure 17b, it can be concluded that the influence of the overvoltage on the GCC control strategy is evidently higher (for the same relative variation) than the influence of the voltage sag. Once more, the known oscillations at twice the grid frequency occur in the injected power (Figure 18a) and positive sequence currents (Figure 18b). Negative sequence currents, with almost identical values for both the *d*- and *q*-axis, are differentiated just by the sign (Figure 18c). After the improved control strategy was applied, the negative sequence currents were successfully mitigated, lessening the oscillations in the injected power and positive sequence currents.

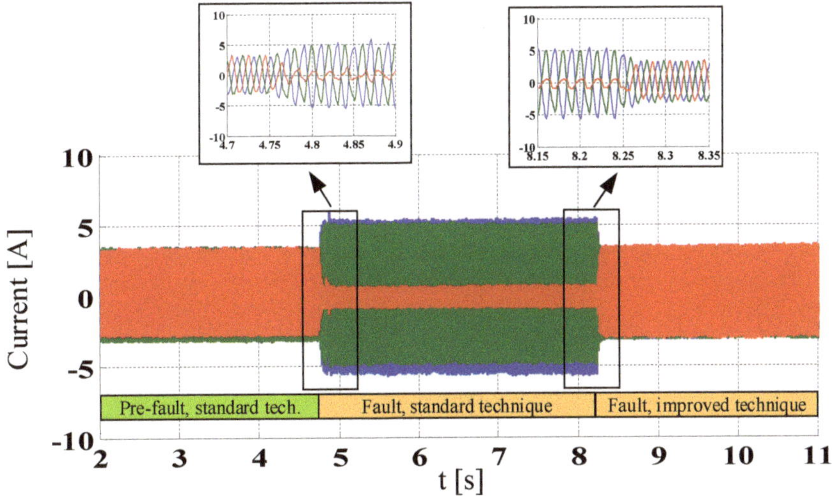

Figure 16. Converter (grid injected) current waveforms at the point of common coupling in one phase for the standard and improved technique—10% overvoltage in one phase.

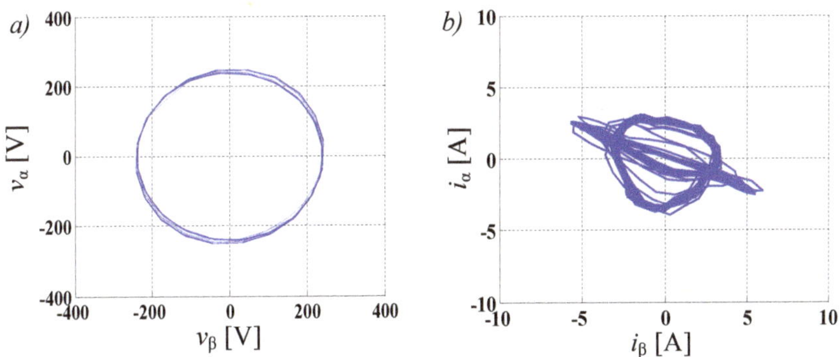

Figure 17. Voltage (**a**) and current (**b**) space vector—10% overvoltage in one phase.

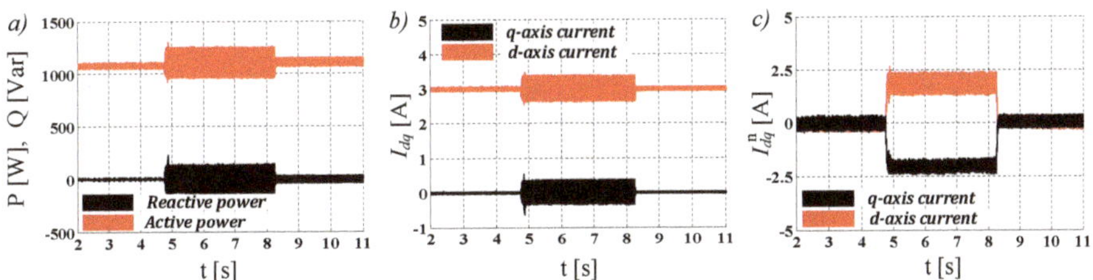

Figure 18. Grid connected converter power (**a**), positive sequence current (**b**), and negative sequence current (**c**) during the experiment—10% overvoltage in one phase.

These experiments clearly demonstrate that the improved control technique can sustain indefinite operation of the GCC under any kind of asymmetrical voltage conditions present at the PCC, even where the standard technique would hit the current protection limitations. In all cases, mitigation of the negative sequence currents led to full controllability of the GCC, with most of the control technique in the positive sequence remaining completely the same.

5. Conclusions

Considering the importance of DER, especially in the present active (and unbalanced) distribution systems, the operation of the GCC under asymmetrical voltage conditions at the PCC will become the most important issue going forward. Since the classical control technique has certain deficiencies when asymmetrical voltages are applied at the GCC input, improvements are necessary in order to fulfill the strict future requirements outlined in the GC. This paper presented an improved control strategy, with relatively simple implementation, aimed at the mitigation of negative sequence currents. The presented strategy maintained the control in the positive sequence at the same level as in the case of the classical strategy, making it a perfect fit for existing grid integrated DERs, as well as for new systems. The improved technique was extensively verified for different voltage sag types (including B, C, and E), arbitrary voltage conditions, as well as for scenarios where asymmetrical overvoltage occurs at the PCC. Not usually tested in the literature, voltage drops with phase shifts and overvoltage can prove to be very demanding scenarios. However, the proposed technique had no problems with dealing with even the most complex grid faults, achieving a great dynamic performance, while, simultaneously, GCC controllability was restored. This offers an extended range of GCC operating states, making it possible for the GCC under asymmetrical voltages to operate indefinitely or according to the relevant GC. Further research will be aimed at introducing additional algorithms to the improved technique in order to offer dynamic grid support capabilities, i.e., special FRT requirements under different voltage asymmetry levels in active unbalanced distribution systems.

Author Contributions: All authors contributed to the research and the paper equally.

Funding: This research received no external funding.

Acknowledgments: This paper is a result of the scientific project No. III 042004 of Integrated and Interdisciplinary Research entitled "Smart Electricity Distribution Grids Based on Distribution Management System and Distributed Generation", funded by Republic of Serbia, Ministry of Education and Science.

Conflicts of Interest: The authors declare no conflict of interest.

References

1. Kamh, M.Z.; Iravani, R. Unbalanced Model and Power-Flow Analysis of Microgrids and Active Distribution Systems. *IEEE Trans. Power Deliv.* **2010**, *25*, 2851–2858. [CrossRef]
2. Twidell, J.; Weir, T. *Renewable Energy Resources*, 2nd ed.; Routledge: London, UK, 2005; ISBN 978-0-419-25330-3.
3. IRENA. *Renewable Energy Prospects for the European Union: Preview for Policy Makers*; IRENA: Abu Dhabi, UAE, 2018.
4. Khushalani, S.; Schulz, N. Unbalanced Distribution Power Flow with Distributed Generation. In Proceedings of the 2005/2006 IEEE/PES Transmission and Distribution Conference and Exhibition, Dallas, TX, USA, 21–24 May 2006; pp. 301–306.
5. Nasr-Azadani, E.; Canizares, C.A.; Olivares, D.E.; Bhattacharya, K. Stability Analysis of Unbalanced Distribution Systems with Synchronous Machine and DFIG Based Distributed Generators. *IEEE Trans. Smart Grid* **2014**, *5*, 2326–2338. [CrossRef]
6. Sharma, I.; Bozchalui, M.C.; Sharma, R. Smart operation of unbalanced distribution systems with PVs and Energy Storage. In Proceedings of the 2013 IEEE International Conference on Smart Energy Grid Engineering (SEGE), Oshawa, ON, Canada, 28–30 August 2013; pp. 1–6.
7. Jia, J.; Yang, G.; Nielsen, A.H. A Review on Grid-connected Converter Control for Short Circuit Power Provision under Grid Unbalanced Faults. *IEEE Trans. Power Deliv.* **2018**, *33*, 649–661. [CrossRef]
8. Ma, K.; Liserre, M.; Blaabjerg, F. Power controllability of three-phase converter with unbalanced AC source. In Proceedings of the 2013 Twenty-Eighth Annual IEEE Applied Power Electronics Conference and Exposition, Long Beach, CA, USA, 17–21 March 2013; pp. 342–350.
9. Ji, Y.; Sun, P.; Wu, Y.; Du, X.; Tai, H.-M.; Gu, S. Power oscillation analysis and control of three-phase grid-connected voltage source converters under unbalanced grid faults. *IET Power Electron.* **2016**, *9*, 2162–2173.

10. Liu, Y.; Li, N.; Fu, Y.; Wang, J.; Ji, Y. Stationary-frame-based generalized control diagram for PWM AC-DC front-end converters with unbalanced grid voltage in renewable energy systems. In Proceedings of the 2015 IEEE Applied Power Electronics Conference and Exposition (APEC), Charlotte, NC, USA, 15–19 March 2015; pp. 678–683.
11. Jiang, W.; Ma, W.; Wang, J.; Wang, L.; Gao, Y. Deadbeat Control Based on Current Predictive Calibration for Grid-Connected Converter Under Unbalanced Grid Voltage. *IEEE Trans. Ind. Electron.* **2017**, *64*, 5479–5491. [CrossRef]
12. Tran, T.; Yoon, S.-J.; Kim, K.-H. An LQR-Based Controller Design for an LCL-Filtered Grid-Connected Inverter in Discrete-Time State-Space under Distorted Grid Environment. *Energies* **2018**, *11*, 2062. [CrossRef]
13. Zhang, W.; Rocabert, J.; Candela, J.I.; Rodriguez, P. Synchronous Power Control of Grid-Connected Power Converters under Asymmetrical Grid Fault. *Energies* **2017**, *10*, 950. [CrossRef]
14. Ozsoy, E.; Padmanaban, S.; Mihet-Popa, L.; Fedák, V.; Ahmad, F.; Akhtar, R.; Sabanovic, A. Control Strategy for a Grid-Connected Inverter under Unbalanced Network Conditions—A Disturbance Observer-Based Decoupled Current Approach. *Energies* **2017**, *10*, 1067. [CrossRef]
15. Yazdani, A.; Iravani, R. A Unified Dynamic Model and Control for the Voltage-Sourced Converter Under Unbalanced Grid Conditions. *IEEE Trans. Power Deliv.* **2006**, *21*, 1620–1629. [CrossRef]
16. Revelo, S.; Silva, C.A. Current reference strategy with explicit negative sequence component for voltage equalization contribution during asymmetric fault ride through: Current Strategy with Negative Sequence Component. *Int. Trans. Electr. Energy Syst.* **2015**, *25*, 3449–3471. [CrossRef]
17. Lai, N.; Kim, K.-H. An Improved Current Control Strategy for a Grid-Connected Inverter under Distorted Grid Conditions. *Energies* **2016**, *9*, 190. [CrossRef]
18. Xiao, P.; Corzine, K.A.; Venayagamoorthy, G.K. Multiple Reference Frame-Based Control of Three-Phase PWM Boost Rectifiers under Unbalanced and Distorted Input Conditions. *IEEE Trans. Power Electron.* **2008**, *23*, 2006–2017. [CrossRef]
19. Galecki, A. Control system of the grid-connected converter based on a state current regulator with oscillatory terms. *Przegled Elektrotechniczny* **2015**, *1*, 67–71. [CrossRef]
20. Bobrowska-Rafal, M.; Rafal, K.; Jasinski, M.; Kazmierkowski, M. Grid synchronization and symmetrical components extraction with PLL algorithm for grid connected power electronic converters—A review. *Bull. Pol. Acad. Sci. Tech. Sci.* **2011**, *59*, 485–497. [CrossRef]
21. Pilo, F.; Jupe, S.; Silvestro, F.; Abbey, C.; Baitch, A.; Bak-Jensen, B.; Carter-Brown, C.; Celli, G.; El Bakari, K.; Fan, M.; et al. *Planning and Optimization Methods for Active Distribution Systems*; CIGRE: Paris, France, 2014.
22. Liu, J.; Gao, H.; Ma, Z.; Li, Y. Review and prospect of active distribution system planning. *J. Mod. Power Syst. Clean Energy* **2015**, *3*, 457–467. [CrossRef]
23. Xu, W.; Dommel, H.W.; Marti, J.R. A generalised three-phase power flow method for the initialisation of EMTP simulations. In Proceedings of the 1998 International Conference on Power System Technology, Beijing, China, 18–21 August 1998; Volume 2, pp. 875–879.
24. Wang, D.T.-C.; Ochoa, L.F.; Harrison, G.P. Modified GA and Data Envelopment Analysis for Multistage Distribution Network Expansion Planning Under Uncertainty. *IEEE Trans. Power Syst.* **2011**, *26*, 897–904. [CrossRef]
25. Dumnic, B.; Popadic, B.; Milicevic, D.; Vukajlovic, N.; Delimar, M. Grid Connected Converter Control Technique in Active Unbalanced Distribution Systems. In Proceedings of the 2018 Mediterranean Conference on Power Generation, Transmission, Distribution and Energy Conversion, Dubrovnik (Cavtat), Croatia, 12–15 November 2018; pp. 1–6.
26. Overview of Power Quality and Power Quality Standards. In *Understanding Power Quality Problems*; IEEE: Piscataway, NI, USA, 2009; ISBN 978-0-470-54684-0.
27. Ignatova, V.; Granjon, P.; Bacha, S.; Dumas, F. Classification and characterization of three phase voltage dips by space vector methodology. In Proceedings of the 2005 International Conference on Future Power Systems, Amsterdam, The Netherlands, 16–18 November 2005; IEEE: Amsterdam, The Netherlands, 2005.
28. Bachschmid, N.; Pennacchi, P.; Vania, A. Diagnostic significance of orbit shape analysis and its application to improve machine fault detection. *J. Braz. Soc. Mech. Sci. Eng.* **2004**, *26*, 200–208. [CrossRef]
29. Popadic, B.; Katic, V.; Dumnic, B.; Milicevic, D.; Corba, Z. Synchronization method for grid integrated battery storage systems during asymmetrical grid faults. *Serb. J. Electr. Eng.* **2017**, *14*, 113–131. [CrossRef]

30. Popadic, B.; Dumnic, B.; Milicevic, D.; Corba, Z.; Vukajlovic, N. Behavior of the Grid Connected Converter in Unbalanced Smart Power Systems. In Proceedings of the 2018 International Symposium on Industrial Electronics (INDEL), Banja Luka, Republic of Srpska, Bosnia and Herzegovina, 1–3 November 2018; IEEE: Banja Luka, Republic of Srpska, Bosnia and Herzegovina, 2018; pp. 1–6.
31. Popadic, B.; Dumnic, B.; Milicevic, D.; Katic, V.; Sljivac, D. Grid-connected converter control during unbalanced grid conditions based on delay signal cancellation. *Int. Trans. Electr. Energy Syst.* **2018**, *12*, e2636. [CrossRef]
32. Blaabjerg, F. (Ed.) *Control of Power Electronic Converters and Systems*; Academic Press: London, UK, 2018; ISBN 978-0-12-805245-7.
33. Vujkov, B.; Dumnic, B.; Popadic, B.; Milicevic, D.; Vukajlovic, N.; Katic, V. Advanced research and development facility for digital control of power electronic based drives. In Proceedings of the 2018 International Symposium on Industrial Electronics (INDEL), Banja Luka, Republic of Srpska, Bosnia and Herzegovina, 1–3 November 2018; IEEE: Banja Luka, Republic of Srpska, Bosnia and Herzegovina, 2018; pp. 1–6.

Article

Modeling and Forecasting Electric Vehicle Consumption Profiles †

Alexis Gerossier *, Robin Girard and George Kariniotakis

MINES ParisTech, PERSEE-Center for Processes, Renewable Energies and Energy Systems, PSL University, 06904 Sophia, Antipolis, France; robin.girard@mines-paristech.fr (R.G.); georges.kariniotakis@mines-paristech.fr (G.K.)

* Correspondence: alexis.gerossier@mines-paristech.fr

† This paper is an extension of the paper Gerossier, A.; Girard, R.; Kariniotakis, G. Modeling Electric Vehicle Consumption Profiles for Short-Term Forecasting and Long-Term Simulation. Presented at the 11th Mediterranean Conference on Power Generation, Transmission, Distribution and Energy Conversion (MEDPOWER 2018), Dubrovnik, Croatia, 12–15 November 2018.

Received: 28 January 2019; Accepted: 22 March 2019; Published: 8 April 2019

Abstract: The growing number of electric vehicles (EV) is challenging the traditional distribution grid with a new set of consumption curves. We employ information from individual meters at charging stations that record the power drawn by an EV at high temporal resolution (i.e., every minute) to analyze and model charging habits. We identify five types of batteries that determine the power an EV draws from the grid and its maximal capacity. In parallel, we identify four main clusters of charging habits. Charging habit models are then used for forecasting at short and long horizons. We start by forecasting day-ahead consumption scenarios for a single EV. By summing scenarios for a fleet of EVs, we obtain probabilistic forecasts of the aggregated load, and observe that our bottom-up approach performs similarly to a machine-learning technique that directly forecasts the aggregated load. Secondly, we assess the expected impact of the additional EVs on the grid by 2030, assuming that future charging habits follow current behavior. Although the overall load logically increases, the shape of the load is marginally modified, showing that the current network seems fairly well-suited to this evolution.

Keywords: electric vehicle; forecasting model; scenario generation; probabilistic evaluation

---

1. Introduction

1.1. Context

The car stock of electric vehicles (EVs)—electric battery and plug-in hybrids—reached 2 million units worldwide in 2016, accounting for 1.1% of the global car market share [1]. This share is expected to rapidly increase over the next 15 years. Charging an EV battery requires a large amount of energy in a small amount of time. In a typical US household, EV charging requires more power than any other appliances (e.g., stoves and dryers) and is solicited just as often (daily or more), see Figure 1. EVs are therefore important appliances to model correctly in order to manage electric household consumption. The increasing number of EVs connected to the grid, coupled with their high power requirement, is challenging the current electrical network with higher overall consumption and additional peaks.

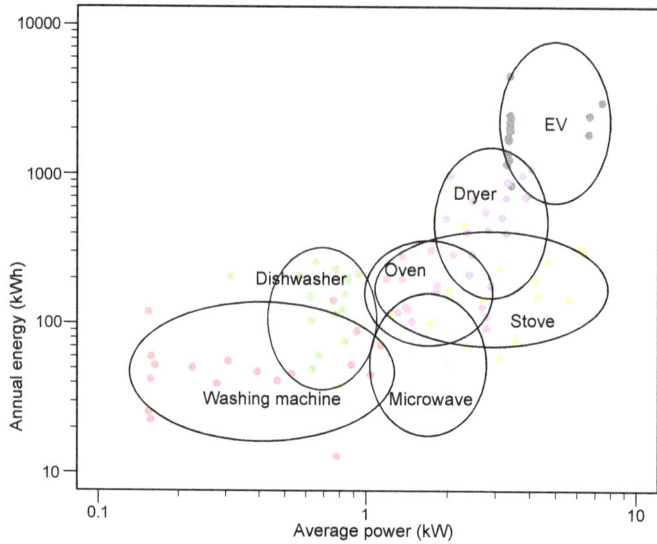

Figure 1. Annual energy (in kWh on y-axis) and average power (in kW on x-axis) of typical appliances of an US household. Source: Pecan Street [2].

The Nordic EV Outlook 2018, published by the Nordic Energy Research [3], gives insight into the EV market in Nordic countries (i.e., Denmark, Finland, Iceland, Norway and Sweden). In particular, the authors provide feedback from the industry in Norway, where the market share of EVs is high (1.9%), pointing out that the electrical grid experiences periodic issues in densely populated urban environments and recreational regions. This is attributed to the number of EVs charging on the grid. The Norwegian energy market regulator suggests that adding an average of 1 kW to the household peak load may result a 4% overloading of the transformers [4]. In Denmark, 20% EV penetration is believed to cause major grid overloading and under-voltage situations [5], while in the UK, a 20% level of penetration is likely to increase the daily peak load by 36% [6].

EVs are used in a multitude of contexts, including professional and leisure usage, meaning that the modeler is faced with a high and challenging variety of charging situations. Due to its nature, an EV can be charged at different places—at home or at work—which rules out a traditional switch-on appliance model. Researchers, such as Bae and Kwasinski, have proposed spatial models to account for different charging stations [7].

Modeling EV charging patterns is a useful tool for several types of study, such as power flow analyses of distribution grids [8], management of smartgrids [9], bottom-up simulations of demand [10], forecasting of charging stations [7], and stabilization of the power system [11]. Furthermore, EV charging involves a controllable load comparable to a washing machine or water heater. As such, EVs offer advantageous flexibility for demand response purposes, for instance, shifting charging cycles when electric demand is low. EV flexibility could be an important input for flexibility models, either at household level [12] or at the aggregated level [13]. Another promising perspective involves injecting the electricity stored in EVs' batteries back into the grid, with so-called "vehicle-to-grid" projects [14].

1.2. Objective

In this study, we use data measured at high-time resolution (i.e., every minute) showing the power drawn from the grid at the charging station. Each charging station is associated with a single privately owned EV. With this data, the charging habits of each EV user are modeled in a probabilistic way. This model is described in Section 2. The charging habits model is then used for forecasting purposes: from short-term, one day ahead, to long term, in 2030.

In Section 3, we generate forecasting scenarios of an EV's consumption profile for the next day. Although our model forecasts a single EV, we validate the scenarios at the aggregated level, i.e., for a fleet of several EVs. We observe that scenarios result in accurate probabilistic forecasts of the fleet's aggregated consumption. In particular, we show that our bottom-up forecasting performs similarly to an advanced machine-learning method that directly forecasts the fleet's aggregated consumption.

In Section 4, we simulate the impact of EVs on the grid in future years. The International Energy Agency (IEA) [15] anticipates a high penetration—around 30%—of EVs in 2030. Employing the four clusters of identified charging habits, we are able to extrapolate the consumption required by a large number of EVs. We show that current charging habits are sufficiently varied so as not to cause major issues on the total electrical load of a region.

1.3. Data Description

A set of 46 privately owned EVs located in Austin, Texas, was selected. Austinites are known to be very climate conscious and supportive of green policies [16], as exemplified by the Pecan Street project run by the University of Texas [2]. The Pecan Street platform provided us with the electric consumption of each EV recorded every minute of the year in 2015. Insight into the households owning the EV was provided; houses were modern (built around 2007) and large (around 195 m^2), meaning that total household consumption was high [17]. In our dataset, electric consumption related to EVs was responsible for approximately 15% of total household consumption.

2. The EV Charging Model

2.1. Processing the EV Time Series

An example of the power drawn by an EV during 36 successive hours is visible in Figure 2. The power drawn was either null, when the EV was not charging, or close to a specific nominal power, when the EV was charging. Based on this this visual inspection, which corresponds to the charging curve measured on a lithium-ion battery by Madrid et al. [18], we modeled a charging period with a block comprising of three parameters, see Figure 3:

1. Nominal power: power demanded from the grid is constant during the whole charging period.
2. Duration of the charging block.
3. Start-up time: moment the day when EV charging starts (between 1 and 60×24).

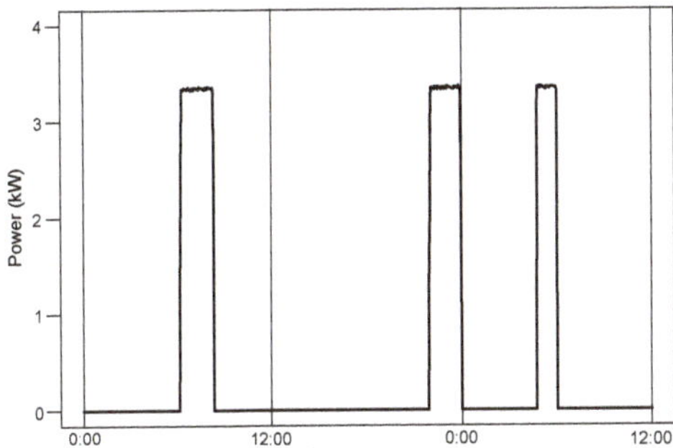

Figure 2. Measured power drawn every minute by an electric vehicle (EV) during 36 successive hours. Power is null when the EV is not charging, and close to a nominal power when charging.

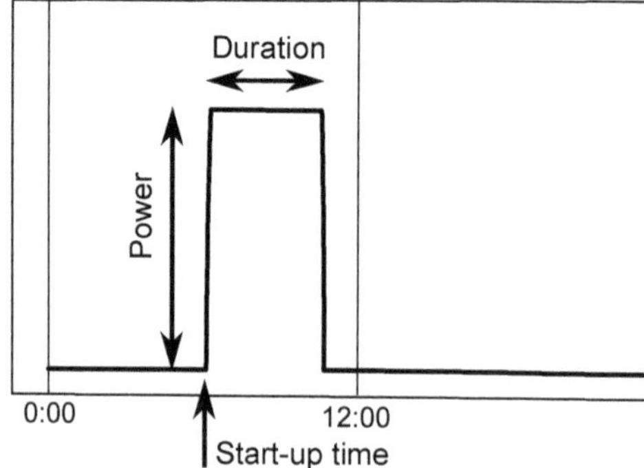

Figure 3. Block representation of the power drawn when charging the battery of a vehicle.

Real measurements did not exhibit perfect blocks. There was a steep ramp up to the nominal power; this ramp usually lasted less than 15 min. The power of the time series fluctuated slightly around a nominal value, translating a noisy phenomenon. This fluctuation was negligible compared to the nominal power, as can be seen on Figure 2. Our hypothesis of perfect charging blocks simplified these two facts.

Our observations indicate that nominal power was always the same for a particular EV as long as there was no technological replacement (i.e., battery and charging station). Such replacements occurred for two of our 46 EVs in the Austin yearlong dataset (nominal power goes from 3.5 to 6.5 kW), requiring a minor adjustment in later modeling. On the other hand, the duration and start-up time were not fixed. Charging blocks almost never started at the exact same time each day, and did not have the same duration; start-up time and duration depend on the unknown users' habits. A realistic depiction of these habits was to describe these two parameters (duration and start-up time) in a probabilistic way. Therefore, an analysis of these parameters was required, meaning that we need to detect charging blocks on the measured power time series.

2.2. Detection of Charging Blocks

We implemented the following procedure to automatically detect the charging blocks of an EV user from the power time series:

1. Detecting nominal power. The density of all of the strictly positive values was estimated, and the maximum of this density function (i.e., the statistical mode) was retrieved as the nominal power.
2. Transforming time series in perfect charging blocks. The raw time series was transformed into a simpler series of two values, either 0 when power was below a threshold fixed at 50% of nominal power, or 1 when it was above.
3. Pre-processing the simple time series. The time series obtained was then refined to account for measurement errors. Missing values were filled in, and any remaining blocks that were too short (less than 20 min) were removed.
4. Detecting duration and start-up time. All timestamps were processed in order to list all durations and associated start-up times in the time series. The day of the year on which the charging blocks occurred was also recorded for forecast applications.

The whole procedure ran fast on an average laptop: less than 30 s for the 525,600 data points of one EV yearly time series.

2.3. Charging Habit Analysis

Once the charging blocks parameters were detected, an analysis of the users' charging habits was possible. An interesting representation was to superimpose every charging block of the year on a graph with the x-axis representing the start-up time, and the y-axis representing the duration of the block. In order to compare users' habits, durations were normalized by the maximal duration observed—so that normalized duration was between 0 and 1. The maximal duration observed translates into the capacity of an EV's battery. Table 1 lists these maximal charging durations, the nominal power of the EVs, and an estimation of the battery capacity. Estimations of capacity matched the battery characteristics provided by manufacturers—such as the 16 kWh battery of a standard EV, or the 60 kWh battery of a premium EV. We also note that the nominal powers used to charge vehicles matched the power outputs of levels 1 and 2, i.e., slow private EV chargers described in the Nordic EV Outlook 2018 [3]. We logically observe no fast chargers ($>$ 22 kW and $\leq$ 150 kW) in our dataset, since these are mostly public and almost negligible compared to slow private chargers [3], although numbers are growing [19].

Table 1. Detected characteristics of electric vehicles (EV).

Nominal Power (kW)	Max Charging Duration (min)	Battery Capacity (kWh)	Number of EVs
1.5	760	19	1
3.5	120	7	2
3.3–3.7	210–240	12–16	35
6.6	150	17	4
6.2–7.4	480–600	53–71	4

From the charging blocks detected, we detailed the number of charging cycles for each day and each EV. Data show that, on average for the 365 days of the year and one EV, there were 150 days with no cycle at all, 158 with only one cycle, and 57 with two or more blocks. Furthermore, considering only days with more than two charging blocks, the main block accounts for more than two thirds of the daily energy requirements. This shows that the main blocks are of paramount importance. Visually, see Figure 4, the characteristics of a main block and any residuals blocks (i.e., second block, third block of the day, etc.) are almost indiscernible. We ascertain this observation with a statistical test comparing the estimated density function in the 2D plan (duration $\times$ start-up time) for the main blocks and residual blocks separately, using package ks available on R software [20]. For more than half of the EVs in the dataset, p-values of the non-parametric test looking for different distribution are below 0.01 [21]. Considering these results and the fact that we observed fewer residual charging blocks, which hinders an accurate statistical model, we considered in the following that main and residual charging blocks came from the same distribution duration $\times$ start-up time.

Similar tests were conducted to determine whether weekdays and weekends follow different patterns. Perhaps surprisingly, for all EVs, we identify no statistical difference (p-value always below 0.01) in charging habits between weekdays and weekends. This is however in agreement with a visual inspection of the charging blocks' characteristics (see Figure 4), where no clear difference stands out. It is also in line with the very low intra-week variations of the electrical load in Texas. However, despite similar habits on weekdays and weekends—EV users charged at the same time and for the same duration—we observed notable differences in the number of times that they charged their EVs each day of the week, e.g., some users almost never charged during the weekend.

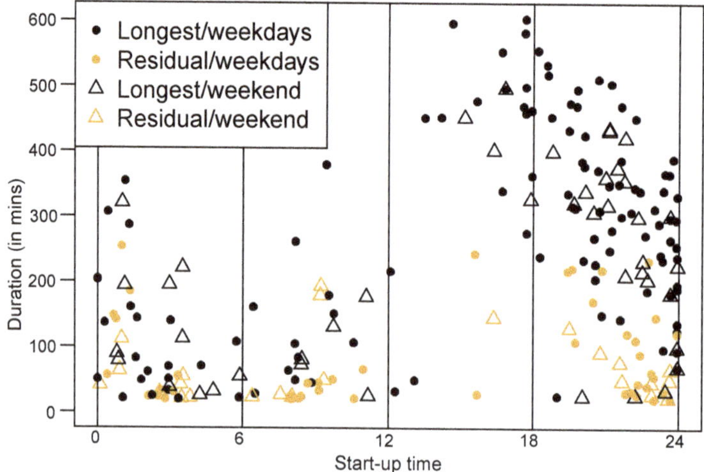

Figure 4. Each point represents a charging block of a specific EV detected during one year. Minute of the start-up time is on the x-axis, and duration of the block on the y-axis. Filled circles, or empty triangles, indicate that the charging occurred during a weekday, or a weekend day, respectively. Colors indicate if this is the longest/main block of the day or a residual block.

2.4. Charging Habits Clustering

By making specific analysis of every user's charging habits, we accurately describe the associated EV's consumption profile. However, to carry out long-term forecasting requires extending these specific habits to a larger scale. We therefore aimed to cluster the charging habits to extract meaningful information that can be extrapolated in a broader context.

First, we estimated the two-dimensional density function for each EV with a kernel density estimator method; a bandwidth matrix common to all EVs was chosen and obtained with a cross-validation method [22]. Then, we compared density functions of two EVs by computing the integrated square differences on the density support. Such values defined proximity between two charging habits. Finally, thanks to a hierarchical clustering based on the Ward linkage method [23], we retrieved four habit clusters from our set of 46 EVs. These clusters represent the charging habits (start-up and duration of charging blocks) regardless of the characteristics of the vehicles, i.e., regardless of the EV's nominal power and total energy capacity.

Table 2 details the four clusters identified, and Figure 5 represents the two-dimensional density functions of each one of the four clusters; the green filled contours represent the density functions estimated with all users in the cluster, and the points denote the charging blocks of one randomly selected user. The first cluster (top-left) gathers the most frequent charging patterns, where EVs are charged during the night and in the morning (52% of the users). This density indicates that most charging cycles occurred before 12:00, and that cycles tended to last longer when started earlier in the night. The second cluster (top-right) gathers users charging in the evening, presumably when people come back from work (20%). The third cluster (bottom-left) gathers users charging throughout the day, but mostly at night-time (20%). The fourth cluster (bottom-right) gathers users charging in the late evening so that their vehicle is charged at a precise moment—such as 03:00 (9%). No statistical link was observed linking the characteristics of the battery and the charging patterns. Although this is in part due to the fact that most batteries are similar, we assume that the two aspects (Tables 1 and 2) are independent.

Table 2. Clusters found.

Cluster	Charging Period	Number of EVs	Frequency
1	Night	24	52%
2	Evening	9	20%
3	Throughout the day	9	20%
4	Late evening	4	9%

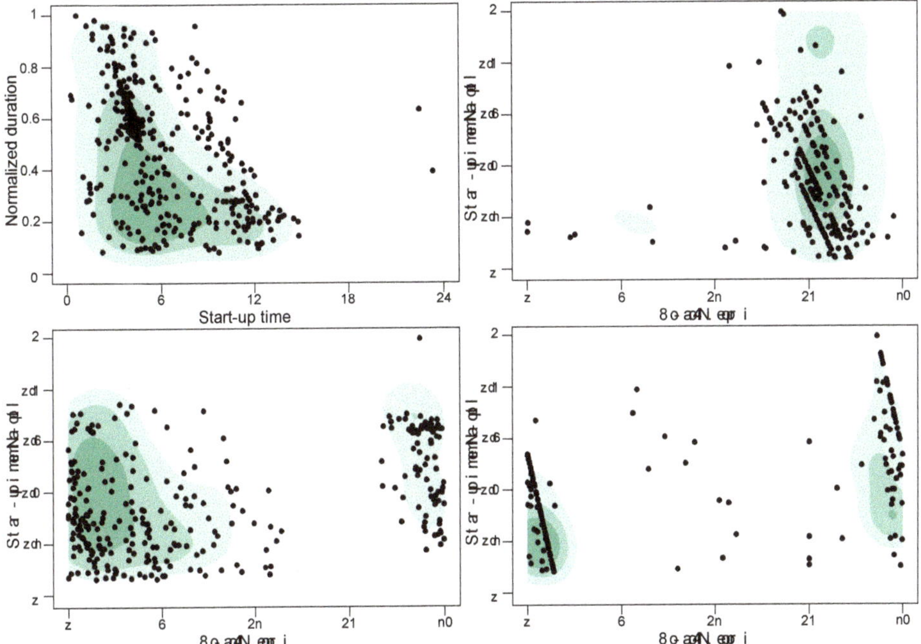

Figure 5. Representation of the charging patterns for the four clusters: the two dimensional density function is estimated with all the EVs of a cluster (green filled contours) and the points represent the charging block a specific EV in the cluster. On the y-axis is the normalized duration, i.e., the charging duration normalized by maximal duration observed for this EV. On the x-axis is the hour of the day.

3. Day-Ahead Forecasting Scenarios of Daily Consumption Profiles

3.1. Scenarios of a Single EV

For a specific EV, charging habits detected from the time series data allow us to forecast daily profile scenarios. This forecasting process is done in three steps:

1. Forecast number of charging blocks for the next day;
2. Forecast possible patterns (normalized duration $\times$ start-up time) for each block;
3. Use of characteristics of the EV (maximal duration and nominal power) to obtain a consumption profile.

For step 1, the forecasting model used is a probabilistic random forest, which provided a convenient way to draw random numbers of charging blocks according to forecast probabilities, using package ranger available on R software [24].

The random forest algorithm has long been established [25] and we detail here the version implemented in the package. The algorithm trains multiple regression trees in parallel, and each tree is fitted on a different dataset. Precisely, one has a training set of K observations $b_1, \ldots, b_K$ (the daily number of charging blocks), and their corresponding inputs sets $s_1, \ldots, s_K$. Each input was made of seven elements: weekday, number of blocks one day ago, number of blocks seven days ago, median number of blocks during the seven previous days, mean temperature of the previous day. These inputs have been selected based on standard inputs for household electricity demand forecasting and empirical tests. A total of J trees, noted $w^1(\cdot), \ldots, w^J(\cdot)$, were to be fitted. These were standard regression trees with manually selected parameters, depth and width, to balance performance and computation time. For every tree, e.g., tree j, a random subset of 80% of the observations was used, so that, for day k, $w^1(s_k), \ldots, w^J(s_k)$ have different values. To make use of the probabilistic aspect of this model, we randomly picked only one value and rounded it to the closest integer. We manually selected a large number of trees, i.e., $J = 10,000$, so as to sufficiently reflect the uncertainty. For step 2, we draw patterns (normalized duration $\times$ start-up time) according to the 2D distribution observed. Forecast charging blocks were drawn from previous ones weighted by a decreasing exponential parameter λ, so that ancient blocks are forgotten. A 2D Gaussian noise, with observed covariance, is added to the block drawn. Checks are operated to rule out impossible situations; overlapping blocks, negative durations and so on.

It is difficult to assess the quality of forecast scenarios for an individual EV. Standard statistical indices (such as mean absolute error) are not adapted to such two-level time series, where the start-up times of charging blocks are highly uncertain. Forecasting methods relying on such indices lead to flat forecasts with no charging block: indeed, a correctly forecast but wrongly timed charging block—e.g., starting at 08:00 instead of 10:00—would be subject to a "double penalty" [26].

3.2. Bottom-Up Forecast of the Aggregated Fleet

Instead of evaluating forecasting performance at the individual level, the aggregated fleet consumption is forecast for the next day with a bottom-up approach. Each EV consumption profile was forecast with the three-step method described in Section 3.1, and the sum of all of the individual scenarios generates a forecast scenario for the aggregated profile. Such a day-ahead forecast is represented in Figure 6 where the aggregated profile can be clearly seen as a sum of the 46 individual EVs profiles.

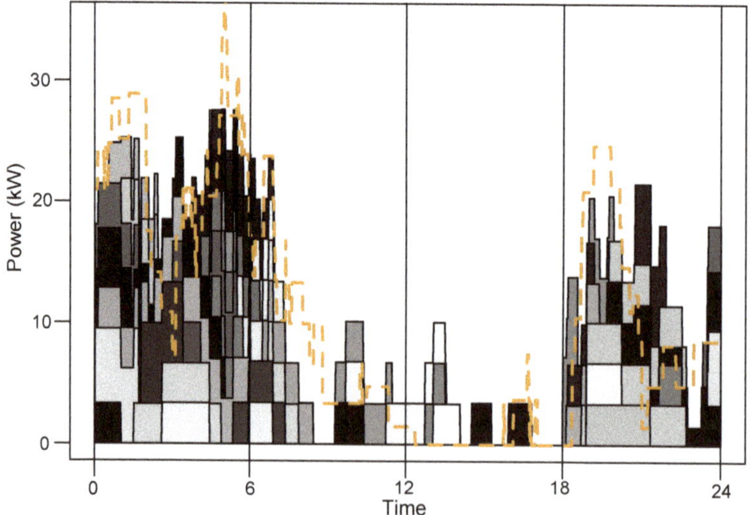

Figure 6. Day-ahead scenario forecast of a fleet of 46 EVs on Saturday 12th December 2015. The orange dashed line shows the actual consumption to be forecast. Each individual scenario is represented by a filled area.

To assess forecasting performance, we generated S scenarios, and turned these scenarios into probabilistic forecasts at each instant by computing quantiles at levels $\tau \in \{0.05, 0.10, ..., 0.95\}$. We compared our method with two benchmarking forecasting models that do not model individual EVs but consider only the aggregated consumption. Contrary to our bottom-up approach that forecast each individual charging before computing the aggregated load of the fleet, these benchmarks directly forecast the aggregated load, and do not consider the individual charges.

First, a persistence model using the value of the aggregated consumption at the same minute on the previous day as a forecast point (no probabilistic framework is proposed with this model). Second, an advanced benchmark is a gradient tree boosting model (GTB), with package gbm available on R software [27]. The gradient tree boosting successively combines weak classifiers to model a complex phenomenon. For regression purposes, the weak classifier is turned into a regression trees of moderate complexity. Therefore, while each tree was quick to compute, the combination of the trees was highly flexible and can model any phenomenon. We detail the algorithm thereafter. One has a training set of T observations $x_1, \ldots, x_T$ and corresponding input sets $i_1, \ldots, i_T$. An input set, e.g., i_t at instant t, was made of five elements: the minute of the day, the weekday, the temperature forecast for the instant, the consumption one day ago, the median consumption during the seven previous days. These are common inputs when forecasting electricity demand [28]. The objective of the GTB is to find a function $g_\tau(i_t)$ that forecast a quantile value at level $\tau \in \{0.05, 0.10, \ldots, 0.95\}$. The level was determined by the choice of the pinball loss function $\mathcal{L}_\tau(x, y)$ defined as

$$\mathcal{L}_\tau(x, y) = (\mathbf{1}(x \leq y) - \tau)(y - x),$$

where $\mathbf{1}(\cdot)$ is the indicator function. In other words, we wanted to find the function such that $\sum_{t=1}^{T} \mathcal{L}_\tau(x_t, g_\tau(i_t))$ was minimal. Finding the optimal function g_τ was made recursively for step $j = 1, \ldots, J$ starting with ${}^{(0)}g_\tau(i_1) = \cdots =^{(0)} g_\tau(i_T) = \text{constant}$. Then, for step j

1. Compute the negative gradient for $t = 1, \ldots, T$

$$^{(j)}z_t = -\frac{\partial}{\partial g(i_t)} \mathcal{L}_\tau(x_t, g_\tau(i_t))\bigg|_{^{(j-1)}g_\tau(i_t)};$$

2. Fit a regression tree ${}^{j}w(\cdot)$ forecasting ${}^{(j)}z_t$;
3. Choose a gradient step

$$\rho^* = \underset{\rho}{\text{argmin}} \sum_{t=1}^{T} \mathcal{L}_\tau(x_t, {}^{(j-1)} g_\tau(i_t) + \rho w(i_t));$$

4. Update estimation, for $t = 1, \ldots, T$

$$^{(j)}g_\tau(i_t) =^{(j-1)} g_\tau(i_t) + \rho w(i_t)).$$

Details of the regression trees, i.e., width and depth, were manually selected in order to optimize performance by the computation time reasonable. This algorithm was in practice slightly altered to improve the training process with a stochastic approach [29], i.e., only 80% of randomly selected observations are used at each step. To avoid overfitting, the number of steps is selected by cross validation.

We evaluated the forecasting quality of the three models with two standard indices: mean absolute error (MAE) for deterministic forecasts and continuous ranked probability score (CRPS) for probabilistic forecasts. Indices were estimated over a training set $\mathcal{T}$ of six months. By noting y_t the actual aggregated load at instant t (to be forecast), and $\hat{y}_t^\tau$ the forecast aggregated load at quantile level τ, then

$$\text{MAE} = \frac{1}{|\mathcal{T}|} \sum_{t \in \mathcal{T}} |y_t - \hat{y}_t^{0.5}|$$

$$\text{CRPS} = \frac{1}{|\mathcal{T}|} \sum_{t \in \mathcal{T}} \int_0^1 2\left(\mathbf{1}\left(y_t \leq \hat{y}_t^\tau\right) - \tau\right)\left(\hat{y}_t^\tau - y_t\right) d\tau,$$

where $\mathbf{1}(\cdot)$ is the indicator function. Note that if only one value is forecast (such as with the persistence model), we consider that this is the value forecast at every quantile level, i.e., the forecast distribution is a Dirac distribution. In this case, the CRPS is equal to the MAE. The obtained scores are reported in Table 3. Thanks to this score, we selected two meta-parameters of our bottom-up forecasts: forgetting parameter $\lambda = 50$ days and number of scenarios $S = 400$. Results show that our bottom-up deterministic forecasts, in addition to the decomposition of the aggregated consumption profile, greatly outperformed the persistence model, and performed similarly to the advanced GTB benchmark. Concerning the probabilistic framework, our bottom-up model was more efficient (i.e., lower CRPS) than GTB. In particular, it was notably more efficient when forecasting the lower tail of the distribution.

Table 3. Indices of two benchmarking methods and our bottom-up approach.

Index	Persistence	GTB	Bottom-Up
MAE (kW)	6.24	4.86	4.87
CRPS (kW)	6.24	3.63	3.59

4. Long-Term Impact of High Penetration of EVs

4.1. Hypotheses

Our dataset described EV charging habits in Austin, Texas. We wanted to extend the study area to a larger region. Therefore, we focused on the South Central region of Texas. The main Texan distribution system operator, Electric Reliability Council Of Texas (ERCOT), defines this region as a weather zone covering 25 contiguous counties, comprising two major cities, Austin and most of San Antonio. According to the Texas Demographic Center, the total population of the 25 counties was 4.8 million in 2017, meaning that there were about 3.4 million vehicles. According to the 2017 National Household Travel Survey, the market share for EVs in Texas at time of writing was around 1.9%, meaning that the current number of EVs—or hybrid EVs—was around 65 thousand in the South Central region.

Considering a 1% immigration scenario in the future, the Texas Demographic Center forecasts that there should be around 6.5 million people by 2030, and thus around 4.6 million vehicles considering that the average number of vehicles per person remains the same. In addition, the IEA's EV30@30 Campaign has set an ambitious goal of a 30% EV penetration rate by 2030. This would result in around 1.4 million EVs by 2030, which means there would be around 1.3 million additional EVs compared to the natural increase of EVs due to population growth over the period. A 30% market share is higher than that anticipated in the detailed study by Musti and Kockelman in 2011 focusing on the city of Austin [30]. These authors estimate the market share to be 19% in 2034 under a favorable feebate scenario.

4.2. Simulation

ERCOT manages electricity representing 90% of the Texan load. The company openly publishes its hourly load curve by weather zone [31]. Without any major technological changes, the load curve should have approximately the same shape in 2030, but at a higher level due to population growth.

For a Tuesday in March, Figure 7 shows the actual load curve in gray, and the expected future load in black.

However, as we estimated, there should be 1.3 million new EVs charging on the grid in 2030, which will impact the load curve. We simulated all of this additional load by generating scenarios for each EV. We considered two possible evolution paths for EVs:

1. Habits and characteristics of EVs remain the same (sample from complete Tables 1 and 2)
2. Habits remain the same but characteristics evolve (sample from last two lines of Table 1 and complete Table 2)

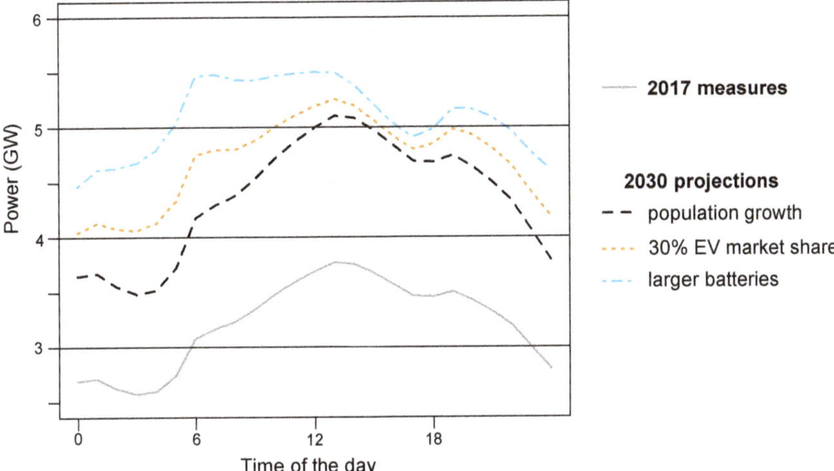

Figure 7. Load curves in South Central zone of Texas in a typical Tuesday in March. The grey solid line represents the load measured in 2017; the black dashed line represents the expected load in 2030 with the same EV market share (1.9%); the orange dotted line represents the expected load with a 30% EV market share and same EV characteristics as of time of writing; the blue mixed line represents this expected load with larger batteries in EVs.

The first evolution path assumes that the habits of future users will fall with the same frequency into the four clusters found in Section 2.4, and that the EVs' characteristics will remain the same. The second path considers the future evolution of EV chargers and batteries. Although fast or ultra-fast chargers are planned to be deployed (nominal power above 22 kW), these are expected to remain public, and public chargers are rarely used compared to private chargers due to consumer preferences. Currently, in the Nordic region, fast chargers represent less than 1% of the total charging load [32], and the growth rate of private chargers is far greater than that of public chargers [3]. However, private chargers may all reach a nominal power of 6.6 kW. We therefore retain only the last two characteristics of Table 1 with half of the future batteries of 17 kWh capacity and half in the 53–71 kWh range capacity. Evolution 1 is represented by the orange line, and evolution 2 by blue line in Figure 7.

Forecast shows that even when a high number of EVs are added to the grid by 2030, their charging only moderately impacts the shape of the load curve at the regional scale of South Central Texas. The overall load is naturally higher with the additional EVs, especially in scenario 2 with larger batteries, but the current charging habits do not cause unmanageable peaks or unstable variability for the load. Both simulations even show that, with the additional EVs, the load curve would be smoothed out during the night, diminishing the intra-day variation. With adequate planning, there should be no major problem with such market share growth. This is in line with other studies assessing the impact of EV charging, such as Luthander's et al in a Swedish case [33]. However, since there could be issues at a local scale, some kind of coordination is required to smart-charge the EVs [6], for instance by optimally scheduling the charging of EV fleets [34], or through targeted price incentives [35].

5. Conclusions

In this paper, we model the consumption profile of EVs from raw power measurements. Based on minute-by-minute power measurements, an algorithm is developed to retrieve each charging block during which an individual charges his or her vehicle. Thanks to this detection, a probabilistic model is proposed to describe the charging habits of the user. From the measurements, we detect five kinds of plugs and EV batteries determining the power drawn from the grid and the battery capacity. Furthermore, we identify four major types of charging habits depending on the duration and start-up time of charging.

Probabilistic models of charging habits can be used to forecast the consumption profiles of single EVs for the next day through scenarios. By adding the scenarios of multiple EVs, models produce bottom-up probabilistic forecasts of the aggregated consumption of a fleet of EVs. A performance evaluation assesses that this method is as efficient as the advanced machine-learning method, but decomposes the aggregated load into single EV consumption profiles.

Since the market share of EVs is expected to greatly increase in the next 15 years, we evaluate the impact of the additional load on the total electrical load in a region in Texas with a population of around 5 million. Based on the four types of charging habits identified on our reduced dataset, we simulate the future load expected in 2030 with and without the EV market share increase, and show that it seems to only moderately impact the shape of the load curve. However, the future of EVs is uncertain, especially concerning the battery capacity and deployment of fast chargers, which may lead to complications for the grid, requiring carefully coordinated charging planning for a large number of vehicles.

Author Contributions: Formal analysis, A.G. and R.G.; Supervision and research methodology, G.K.; Writing—original draft, A.G.; Writing—review & editing, R.G. and G.K.

Funding: This research was funded by the European Union under the Horizon 2020 Framework Programme grand agreement No. 645963 as part of the research and innovation project SENSIBLE (Storage ENabled SustaInable energy for BuiLdings and communitiEs—www.projectsensible.eu).

Acknowledgments: The authors would like to thank Pecan Street Inc. for the provision of power data used in this study.

Conflicts of Interest: The authors declare no conflict of interest.

Abbreviations

The following abbreviations are used in this manuscript:

CRPS	Continuous Ranked Probability Score
ERCOT	Electric Reliability Council Of Texas
EV	Electric Vehicle
GTB	Gradient Tree Boosting
IEA	International Energy Agency
MAE	Mean Absolute Error
UK	United Kingdom
US	United States

References

1. International Energy Agency. *Global EV Outlook 2017*; IEA Publishing: Paris, France, 2017.
2. Pecan Street Inc. Dataport. 2018. Available online: https://dataport.cloud/ (accessed on 24 January 2019).
3. International Energy Agency. *Nordic EV Outlook 2018*; IEA Publishing: Paris, France, 2018.
4. *Status of NVE's Work on Network Tariffs in the Electricity Distribution System*; The Norwegian Water Resources and Energy Directorate (NVE): Oslo, Norway, 2016.

5. Wu, Q.; Jensen, J.M.; Hansen, L.H.; Bjerre, A.; Nielsen, A.H.; Østergaard, J. *EV Portfolio Management and Grid Impact Study*; Annual Report; Technical University of Denmark: Lyngby, Denmark, 2009.
6. Qian, K.; Zhou, C.; Allan, M.; Yuan, Y. Modeling of load demand due to EV battery charging in distribution systems. *IEEE Trans. Power Syst.* **2011**, *26*, 802–810. [CrossRef]
7. Bae, S.; Kwasinski, A. Spatial and temporal model of electric vehicle charging demand. *IEEE Trans. Smart Grid* **2012**, *3*, 394–403. [CrossRef]
8. Putrus, G.; Suwanapingkarl, P.; Johnston, D.; Bentley, E.; Narayana, M. Impact of electric vehicles on power distribution networks. In Proceedings of the 2009 IEEE Vehicle Power and Propulsion Conference, Dearborn, MI, USA, 7–10 September 2009; pp. 827–831.
9. Mwasilu, F.; Justo, J.J.; Kim, E.K.; Do, T.D.; Jung, J.W. Electric vehicles and smart grid interaction: A review on vehicle to grid and renewable energy sources integration. *Renew. Sustain. Energy Rev.* **2014**, *34*, 501–516. [CrossRef]
10. Dickert, J.; Schegner, P. Residential load models for network planning purposes. Modern Electric Power Systems (MEPS). In Proceedings of the 2010 IEEE International Symposium, Wroclaw, Poland, 20–22 September 2010; pp. 1–6.
11. Tomić, J.; Kempton, W. Using fleets of electric-drive vehicles for grid support. *J. Power Sources* **2007**, *168*, 459–468. [CrossRef]
12. Correa-Florez, C.A.; Gerossier, A.; Michiorri, A.; Girard, R.; Kariniotakis, G. Residential electrical and thermal storage optimisation in a market environment. *CIRED-Open Access Proc. J.* **2017**, *2017*, 1967–1970. [CrossRef]
13. Ponocko, J.; Milanovic, J.V. Forecasting Demand Flexibility of Aggregated Residential Load Using Smart Meter Data. *IEEE Trans. Power Syst.* **2018**, *33*, 5446–5455. [CrossRef]
14. Gough, R.; Dickerson, C.; Rowley, P.; Walsh, C. Vehicle-to-grid feasibility: A techno-economic analysis of EV-based energy storage. *Appl. Energy* **2017**, *192*, 12–23. [CrossRef]
15. International Energy Agency. *Energy Technology Perspective 2017*; IEA Publishing: Paris, France, 2017.
16. Smith, C.; Fowler, M.; Greene, E.; Nielson, C. Carbon emissions and climate change: A study of attitudes and their relationship with travel behavior. In Proceedings of the Transportation Research Board's National Transportation Planning Applications Conference, Houston, TX, USA, 17–21 May 2009.
17. Beckel, C.; Sadamori, L.; Staake, T.; Santini, S. Revealing household characteristics from smart meter data. *Energy* **2014**, *78*, 397–410. [CrossRef]
18. Madrid, C.; Argueta, J.; Smith, J. *Performance Characterization—1999 Nissan Altra-EV with Lithium-ion Battery*; Southern California EDISON: Rosemead, CA, USA, 1999.
19. Ionity. *Fast Charging Station Network Starts to Take Shape: Site Partners for 18 European Countries Secured*; Emobility plus: Navi Mumbai, India, 2017.
20. Duong, T. *k: Kernel Smoothing*; R Package Version 1.11.1; R core team: Kingston, ON, Canada, 2018.
21. Anderson, N.H.; Hall, P.; Titterington, D.M. Two-sample test statistics for measuring discrepancies between two multivariate probability density functions using kernel-based density estimates. *J. Multivar. Anal.* **1994**, *50*, 41–54. [CrossRef]
22. Duong, T.; Hazelton, M.L. Cross-validation bandwidth matrices for multivariate kernel density estimation. *Scand. J. Stat.* **2005**, *32*, 485–506. [CrossRef]
23. Murtagh, F.; Legendre, P. Ward's hierarchical agglomerative clustering method: Which algorithms implement Ward's criterion? *J. Classif.* **2014**, *31*, 274–295. [CrossRef]
24. Wright, M.N.; Ziegler, A. ranger: A Fast Implementation of Random Forests for High Dimensional Data in C++ and R. *J. Stat. Softw.* **2017**, *77*, 1–17. [CrossRef]
25. Breiman, L. Random forests. *Mach. Learn.* **2001**, *45*, 5–32. [CrossRef]
26. Haben, S.; Ward, J.; Greetham, D.V.; Singleton, C.; Grindrod, P. A new error measure for forecasts of household-level, high resolution electrical energy consumption. *Int. J. Forecast.* **2014**, *30*, 246–256. [CrossRef]
27. Ridgeway, G. *gbm: Generalized Boosted Regression Models*; R package version 2.1.3; R core team: Kingston, ON, Canada, 2017.
28. Gerossier, A.; Girard, R.; Kariniotakis, G.; Michiorri, A. Probabilistic day-ahead forecasting of household electricity demand. *CIRED-Open Access Proc. J.* **2017**, *2017*, 2500–2504. [CrossRef]
29. Friedman, J.H. Stochastic gradient boosting. *Comput. Stat. Data Anal.* **2002**, *38*, 367–378. [CrossRef]

30. Musti, S.; Kockelman, K.M. Evolution of the household vehicle fleet: Anticipating fleet composition, PHEV adoption and GHG emissions in Austin, Texas. *Transp. Res. Part A Policy Pract.* **2011**, *45*, 707–720. [CrossRef]
31. ERCOT. ERCOT Load History. Available online: http://www.ercot.com/gridinfo/load/load_hist/ (accessed on 11 March 2019).
32. *Nordic EV Barometer 2018*; Norsk Elbilforening (Nordic Electric Vehicle Association): Oslo, Norway, 2018.
33. Luthander, R.; Shepero, M.; Munkhammar, J.; Widén, J. Photovoltaics and opportunistic electric vehicle charging in the power system—A case study on a Swedish distribution grid. In Proceedings of the 7th International Workshop on Integration of Solar into Power Systems, Berlin, Germany, 24–25 October 2017.
34. He, Y.; Venkatesh, B.; Guan, L. Optimal scheduling for charging and discharging of electric vehicles. *IEEE Trans. Smart Grid* **2012**, *3*, 1095–1105. [CrossRef]
35. Cao, Y.; Tang, S.; Li, C.; Zhang, P.; Tan, Y.; Zhang, Z.; Li, J. An optimized EV charging model considering TOU price and SOC curve. *IEEE Trans. Smart Grid* **2012**, *3*, 388–393. [CrossRef]

Article

Evolutionary Multi-Objective Cost and Privacy Driven Load Morphing in Smart Electricity Grid Partition

Miltiadis Alamaniotis * and Nikolaos Gatsis

Department of Electrical and Computer Engineering, University of Texas at San Antonio, San Antonio, TX 78201, USA

* Correspondence: miltos.alamaniotis@utsa.edu; Tel.: +1-210-458-6237

Received: 8 April 2019; Accepted: 21 June 2019; Published: 26 June 2019

Abstract: Utilization of digital connectivity tools is the driving force behind the transformation of the power distribution system into a smart grid. This paper places itself in the smart grid domain where consumers exploit digital connectivity to form partitions within the grid. Every partition, which is independent but connected to the grid, has a set of goals associated with the consumption of electric energy. In this work, we consider that each partition aims at morphing the initial anticipated partition consumption in order to concurrently minimize the cost of consumption and ensure the privacy of its consumers. These goals are formulated as two objectives functions, i.e., a single objective for each goal, and subsequently determining a multi-objective problem. The solution to the problem is sought via an evolutionary algorithm, and more specifically, the non-dominated sorting genetic algorithm-II (NSGA-II). NSGA-II is able to locate an optimal solution by utilizing the Pareto optimality theory. The proposed load morphing methodology is tested on a set of real-world smart meter data put together to comprise partitions of various numbers of consumers. Results demonstrate the efficiency of the proposed morphing methodology as a mechanism to attain low cost and privacy for the overall grid partition.

Keywords: load morphing; NSGA-II; smart grid; grid partition; multi-objective optimization; Pareto theory

1. Introduction

The introduction of communication and information technologies in the current power infrastructure has accommodated the digital connection of electricity market participants [1] and has enabled their active participation in various activities pertained to grid operation. The concept of an energy Internet as introduced in [2] is a characteristic example of the integration of digital connectivity with power systems. Through active participation, consumers play a significant role in determining the shape of the final load demand, mainly by responding to price signals with their individual demand [3]. In addition, consumers are now exposed to a high volume of information that is utilized for making optimal decisions and fulfilling their goals. Notably, the goals of electricity consumers encompass the maximum fulfillment of energy needs and the minimum possible cost [4].

On one hand, the introduction of digital connectivity in the power grid has leveraged the role of consumers resulting in the enhanced grid stability, minimization of power losses and decreased cost of operation [5]. On the other hand, it came with a reduced degree of consumer privacy given that information sharing and exchange may reveal details about the private life of the consumers. More specifically, load profiles, if shared, can be used to infer the consuming activities of a specific customer. For instance, the use of a washer may be inferred by identifying the consumption pattern of the washer in the load profile. Such information may be utilized by third parties for advertising reasons or for nefarious purposes, where burglars may infer the presence or not of the owner and break into the house. Therefore, shared load information comprises of a point of vulnerability that may be used to compromise privacy of consumers in the smart cities of the future [6].

In a market, it is anticipated that consumers are interested in purchasing products of the highest quality at the lowest possible cost. With regard to electricity consumption, electricity consumers care about satisfying their maximum demand while attaining the lowest cost; in other words, they would like to minimize their electricity bill, and fully perform their planned activities [4]. In general, the decision-making process of consumers, including electricity consumers, is a cost driven approach, where the consumer has as a first priority the minimization of the overall cost [7]. In that case, the consumer is prone to morphing his/her load demand in order to retain the cost at a comfortable level. In smart grids, load demand morphing refers to the actions of either cancelling, or shifting the load demand [3] with respect to an initial plan. Cancelling load refers to abandon the scheduling of specific consuming actions, while shifting refers to postponing the consuming actions at a later time, usually at times where the price is lower [8]. The response of consumers with their load demand to prices set by the market operator is known with the general term of "demand response" [9].

There are several approaches that deal with the demand response of consumers aiming at minimizing the cost of consumption. For instance, in [10] an approach that optimizes the electric appliances scheduling for demand response is presented, while in [11] an approach that reduces the load variation limits to minimize consumption costs is introduced. The concept of Virtual Budget as an efficient method for optimizing the electricity cost of demand scheduling using anticipation is proposed in [3], while a sliding window driven method, which utilizes streamed big data for real time electricity consumption optimal adjustment, is proposed and tested in [12]. Furthermore, an optimization algorithm for residential consumption pattern flattening by identifying the time-of-use tariff that minimizes the overall consumption cost is introduced in [13]. In [14], methods that assume the use of interruptible tasks are proposed, whereas in [15] an informatics solution that is based on the synergism of three models in optimizing household appliance management. Furthermore, an autonomous system for demand response that achieves minimal cost for participating in demand response programs consumers is introduced in [16], and a similar approach that assumes consumers response to utility signals prior to any decision making is presented in [17]. Moreover, an approach for scheduling the electricity consumption of a residential community based on the aggregated payment is introduced in [18]. A coordinated approach considering a community of prosumers is discussed is in [19], while the distributed coordination of a grouped consumption using the alternative directions method of multipliers is introduced in [20]. Overall, the variety of demand response approaches aim at securing the operation of the grid [21], while the consumers target to minimize their consumption expenses [22]. However, there are demand response methods that rather focus on consumers' privacy preserving than the cost-minimizing. An example is presented in [23], where a privacy preserving method employs data encryption in the form of a homomorphic encryption of the group aggregated demand. Further, a method focusing on incentive-based demand response of consumers using cryptographic primitives, such as identity-committable signatures and partially blind signatures, is presented and tested in [24]. Data exchange architectures to ensure the privacy of electricity consumption profiles are presented in [25] and [26], where in the first case trusted-platform modules for advanced metering infrastructure (AMI) are used, and in the second case a set of Internet-of-Things tools are put together. In [27], a secure algorithm tailored for bidding driven markets is proposed utilizing cryptographic primitives without any explicit third trusted party. Lastly in [28], a study that involves the use of privacy threat models together with attributed based encryption is provided.

The proposed demand response methods are focusing on single residents and are proposed within the framework of optimizing the power grid infrastructure. Furthermore, they do not exploit the ubiquitous digital connectivity that will be the backbone of the smart cities and smart grids. In this work, we propose an approach that builds upon that connectivity driven future. Notably, we see that both the digital connectivity and the smart grid technologies are the enablers for smart cities.

In particular, we assume that consumers, which are also residents of a smart city, are able to connect and form "virtually connected groups" [29]. This grouping of the residents/consumers is performed at the cyber level and subsequently allows consumers to form energy partitions within the smart grid [30]. Hence, we will exploit digital connectivity of smart grids to allow consumers within the same partition to collaborate and formulate the partition's demand response. The partition demand response is a single load pattern that coincides with the morphed aggregation of the individual demand patterns of the consumers within the partition, an idea that has been proposed in [31].

In this work, we push further the envelope in load morphing by assuming that the partition demand response results from the concurrent consideration of the partitions consumption expenses and privacy. The paper introduces a new approach where citizens collaborate in order to concurrently attain low cost and high privacy as a group. A premature version of this method was presented in [32], where its high potential for smart grids was highlighted. As compared to [32], this paper presents in more details the morphing method, while it applies it in a higher variety of smart meter data taken from the power grid of Ireland [33]. Notably, the proposed approach considers the morphing problem as a multi-objective problem whose solution is located by an evolutionary algorithm. In this work, the non-dominated sorting genetic algorithm-II (NSGA-II) is adopted to provide a solution to the final morphing of the demand [34]. NSGA-II is a genetic algorithm that utilizes the Pareto optimality theory to identify a solution that optimizes both the cost and privacy of the grid partition [35].

At this point it should be noted, that the current work significantly differs from the works in [29–31]. Those works focus only on enhancing the privacy of the group of the consumers without considering the cost of the final morphed consumption pattern, whereas the current work also explicitly considers the cost of consumption. Furthermore, in References [29,31] a genetic algorithm is adopted to solve a single objective optimization problem, while in the current work the genetic algorithm is selected for locating a solution to a multi-objective problem.

The innovation in the current work is the concurrent consideration of both privacy and consumption cost in a grid partition, and their concurrent optimization through Pareto optimality, which finds the optimal tradeoff between the two objectives, i.e., cost and privacy. This work aspires to show that the morphing of partition consumption driven by those two objectives is a complex optimization problem with multiple constraints, that can be solved by genetic algorithms. Notably, genetic algorithms have the ability to always identify a global optimal solution (or near optimal) to complex problems independently of the number of objectives and constraints.

The roadmap of this paper is as follows: In the next section a brief introduction to evolutionary computing and Pareto optimality is given, while in Section 3 the morphing methodology is presented. In Section 4, the results on a set of real-world datasets are presented and discussed, while Section 5 concludes the paper.

2. Background

2.1. Introduction to Evolutionary Computing

The set of stochastic optimization techniques whose framework has been inspired by various natural evolution processes are known as "evolutionary computing". Notably, evolutionary computing algorithms are identified as part of the broader class of artificial intelligence algorithms. Some of the most widely used evolutionary algorithms, which have been utilized in various applications, encompass genetic algorithms, genetic programming, swarm intelligence, and artificial immune systems [36].

The main principle upon which evolutionary computing algorithms are built is the selection of an initial population and the subsequent rise of the population fitness by following a natural selection process, based upon the survival of the fittest individuals. Similar to nature, the evolutionary computing algorithms select as the final solution the fittest population individuals. In sum, the essential steps that evolutionary algorithms have in common are the following [36]:

- Maintenance of a set of candidate solutions (population);
- Fitness evaluation and sampling of the current population of solutions; and
- Recombination and mutation of solutions to generate new improved solutions.

The above steps drive the implementation of various evolutionary algorithms despite the fact that each one differs in technical details. The outline of a general evolutionary algorithm is depicted in Figure 1.

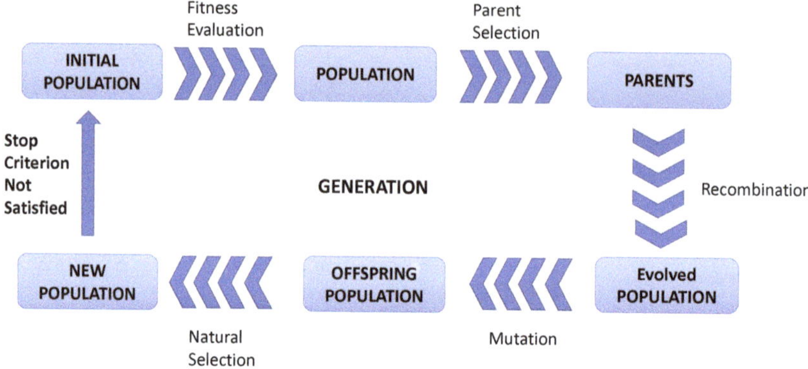

Figure 1. Outline of a general evolutionary computing algorithm.

It should be noted that the full cycle as shown in Figure 1 denotes a generation; thus, the generation is an algorithmic iteration, where transition to a new iteration is initiated when the stopping criterion is not satisfied.

Genetic algorithms are evolutionary algorithms that are used to find solutions in optimization problems that are either single-objective or multi-objective. They are powerful algorithms that identify an optimal solution or a near optimal solution. In addition, they are able to provide solutions to highly complex problems where conventional convex optimization algorithms fail. However, their operation needs the initial determination of a set of variables, i.e., initial population, probability of mutation etc, that may affect their performance. Furthermore, genetic algorithms cover a high area of the search space, but seeking for a solution in a wide area makes them computationally slow. However, their ability of find a global solution promotes them as an attractive option, despite their slow execution time. In this text, the non-dominated sorting algorithm-II will be used for identifying a solution by utilizing the Pareto optimality theory.

2.2. Pareto Optimality

Problems that involve the optimization of multiple objectives are called multi-objective optimization problems, or alternatively vector optimization. The multi-objective problems are following general formulation for minimization problems:

$$\begin{aligned} \min \mathbf{C}(\mathbf{x}) &= [C_1(\mathbf{x}), C_2(\mathbf{x}), \ldots, C_N(\mathbf{x})] \\ s.t.\ \mathrm{f}_i(\mathbf{x}) &\leq 0, \quad i = 1, \ldots, k \\ \mathrm{g}_j(\mathbf{x}) &= 0, \quad j = 1, \ldots, m \end{aligned} \tag{1}$$

with $C_{\#}()$ denoting an objective function, N being the population of objective functions, k the population of inequality constraints, m the number of equality constraints, and $\mathrm{f}_{\#}()$ and $\mathrm{g}_{\#}()$ being valid analytical functions.

As opposed to the single-objective optimization, multi-objective optimization problems typically do have more than a single global solution. As a result, there is a need to determine a set of criteria as to whether a solution is recognized as optimal or not. To that end, the set of criteria that identify a solution as an optimal one are set by the *Pareto optimality* theory [37]. According to Pareto theory, "*a point*, $\mathbf{x}^* \in \mathbf{X}$, *is Pareto optimal iff there does not exist another point*, $\mathbf{x} \in \mathbf{X}$, *such that* $\mathbf{C}(\mathbf{x}) \leq \mathbf{C}(\mathbf{x}^*)$, *and* $C_i(\mathbf{x}) \leq C_i(\mathbf{x}^*)$ *for at least one function* [38]". To make it clearer, a solution is Pareto optimal if there is no other solution that improves at least one objective without adversely affecting any of the rest objectives.

Notably, the population of Pareto solutions pertained to a specific problem can be infinite. The population of the Pareto solutions consist of the *Pareto optimal set*, while the projection of the Pareto optimal set on the objective space is known as the *Pareto frontier* [38]. A general illustrative example of Pareto frontier consisting of two objectives is depicted in Figure 2.

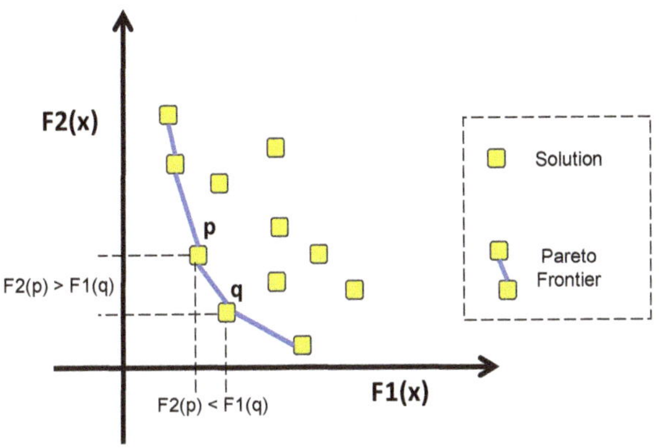

Figure 2. Illustrative example of a Pareto Frontier for a two-objective problem [32].

In addition to the Pareto theory, the concept of a *non-dominated* or *dominated solution* is of significance. A solution $\mathbf{x}_1$ is identified as dominated over $\mathbf{x}_2$ when: (i) The solution $\mathbf{x}_1$ is not worse than solution $\mathbf{x}_2$ in all objective functions, and (ii) the solution $\mathbf{x}_1$ is strictly better than $\mathbf{x}_2$ in at least one objective function [38]. If the above conditions are not valid at the same time, then the solution $\mathbf{x}_1$ is non-dominated over $\mathbf{x}_2$ [38].

3. Grid Partition Load Morphing Methodology

In this section, the load morphing of smart grid partitions methodology is presented. It should be noted that the way the partitions are formed in the grid is not of interest to the current manuscript. For instance, one way to form partitions is via grid partitioning as discussed in [39]. The methodology presented in this manuscript formulates a multi-objective problem composed of a dual of objective functions [8]. The two objectives, which express the consumption cost and the degree of privacy in the grid partition respectively, are minimized by seeking an optimal solution to a multi-objective problem. The NSGA-II algorithm is applied to identify an optimal solution. The block diagram of the proposed methodology is presented in Figure 3, where its individual steps are clearly provided.

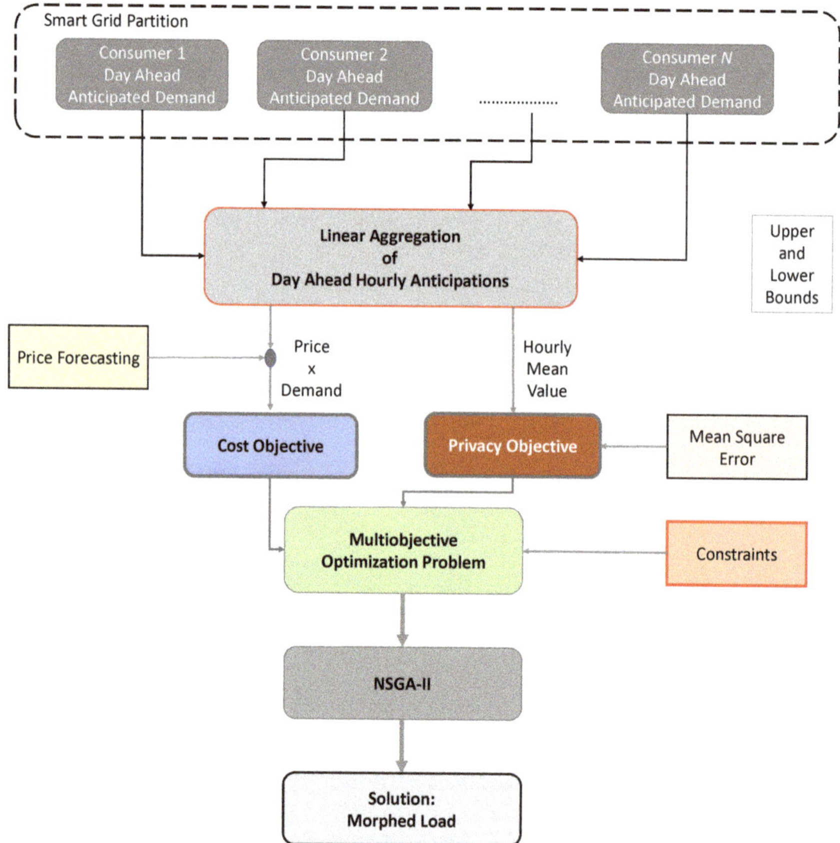

Figure 3. Block diagram of the grid partition load morphing methodology.

Initially, the consumers within the grid partition anticipate their day ahead load demand. The anticipation is conducted at an hourly level, i.e., that means a set of 24 load values for the next day is provided. The overall load anticipation for a whole day is denoted as given below:

$$\Sigma L = \sum_{i=1}^{24} L_i \tag{2}$$

where L_i is the anticipated load at hour i. In addition to the anticipated load, the consumers are also providing their hourly upper and lower bounds of their anticipated load denoted as U_i and F_i for the hour i. These bounds coincide with the maximum and minimum values between which the initial anticipation can be morphed for a specific hour, and we call them the "morphing bounds". To make it clearer, when the anticipation for the hour i is L_i, then the anticipation can be morphed within the tube $[L_i - F_i, L_i + U_i]$.

In the next step, the individual consumers' anticipated loads are aggregated and a single anticipated load demand signal is obtained. Thus, the aggregated pattern expresses the anticipated load of all the partition consumers and is expressed as:

$$A_i = \sum_{j=1}^{N} L_j^i \quad i = 1, \ldots, 24 \tag{3}$$

where L_j^i is the load of consumer j at hour i, and N is the population of consumers. Likewise, the individual morphing bounds are also aggregated providing the respective aggregation morphing bounds:

$$U_i^A = \sum_{j=1}^{N} U_j^i \quad i = 1, \ldots, 24 \tag{4}$$

$$F_i^A = \sum_{j=1}^{N} F_j^i \quad i = 1, \ldots, 24 \tag{5}$$

where U_i^A and F_i^A are the upper and lower aggregation morphing bounds for the hour i. At this point, we introduce the morphed aggregated load, which is denoted as:

$$A_i^M = \alpha_i \sum_{j=1}^{N} L_j^i \quad i = 1, \ldots, 24 \tag{6}$$

with A_i^M being the morphed aggregated load at hour i, and α_i the morphing factors for the hour i. The morphing factors express the degree of morphing, where: (i) $\alpha_i = 1$ denotes no morphing, (ii) $\alpha_i < 1$ denotes decreasing of the initial anticipation, and (iii) $\alpha_i > 1$ denotes increasing of the initial anticipation. Notably, for $\alpha_i = 1$ Equation (6) drops down to Equation (3).

Once the aggregated values have been computed, then the objective functions are formulated. According to Figure 3, two objectives are formulated, namely, the cost and privacy objectives. The cost objective expresses the daily cost of purchasing the anticipated load and is formulated as:

$$\text{Cost} = \sum_{i=1}^{24} A_i^M \cdot P_i \tag{7}$$

with P_i being the day ahead forecasted electricity price for the hour i.

Formulation of the second objective, which expresses the degree of privacy, is more complex than that of the cost. It should be noted that as a measure of privacy we assume the degree of variance in the load pattern. On one hand a highly varying pattern is a carrier of information that can be easily extracted and subsequently lead to inferences about consumption activities. In other words, variability can be a source of information–the peaks and valleys of the pattern can be associated with consumption activities. On the other hand, a constant load pattern exhibits no variance and thus, inference making becomes challenging. In this work, we aim at deriving a constant load pattern by morphing the aggregated demand. To that end, we adopt a target constant pattern that is equal to the mean value of the aggregated value given below:

$$M = \frac{1}{24} \sum_{i=1}^{24} A_i \tag{8}$$

where A_i is the aggregated value for hour i computed by Equation (3). Furthermore, to quantify the degree of difference between the aggregated load and the mean aggregated value, an error measure is adopted. More specifically, the mean square error is utilized as the objective for expressing the degree of privacy as the distance of the aggregated demand to the mean value. Thus, the privacy objective takes the following form:

$$\text{Privacy} = \frac{1}{24} \sum_{i=1}^{24} \left(A_i^M - M\right)^2 \tag{9}$$

where M is taken by Equation (8) and A_i by Equation (2).

The objective functions as defined by Equations (7) and (9) are accompanied with a set of constraints. The constraints, which are in the form of box constraints, express the morphing bounds of the aggregated load and are formulated as:

$$F_i^A \le A_i^M \le U_i^A \quad i = 1, \ldots, 24 \tag{10}$$

with A_i^M being the morphed aggregated load (see Equation (6)). As shown in Equation (10), there are 24 box constraints, i.e., one for every hour of the day.

At this point, the two objectives and the respective constraints have been fully defined and therefore, we are able to define the multi-objective problem utilized for load morphing of the grid partition. The particular form of the multi-objective problem is given below:

$$\left\{ \begin{array}{l} \underset{\alpha_i}{\text{minimize}}[\text{Cost}, \text{Privacy}] \\ \text{w.r.t. } F_i^A \le A_i^M \le U_i^A \\ \qquad i = 1, \ldots, 24 \\ \text{where Cost} = \sum\limits_{i=1}^{24} A_i^M \cdot P_i \\ \qquad \text{Privacy} = \frac{1}{24} \sum\limits_{i=1}^{24} \left(A_i^M - M\right)^2 \\ \qquad A_i^M = \alpha_i \sum\limits_{j=1}^{N} L_j^i \end{array} \right\} \tag{11}$$

where the optimization process takes the form of a minimization of the two objectives.

The multi-objective minimization problem in Equation (11) is solved utilizing evolutionary computing and more specifically the NSGA-II algorithm. The NSGA-II seeks for non-dominated solutions, which satisfy by default the Pareto optimality criterion. The identified solution which is comprised of a set of 24 optimal morphed values, i.e., optimal α_i^{opt}, $i = 1, \ldots, 24$, is the final solution of the problem. Each morphed factor expresses the degree of morphing of the load for the hour i. Having computed the morphed values, then the final morphed pattern is obtained by plugging-in the identified solution to Equation (6):

$$A_i^M = \alpha_i^{opt} \sum_{j=1}^{N} L_j^i \quad i = 1, \ldots, 24 \tag{12}$$

where the superposition of the morphed values provides the final load curve of the smart grid partition. To conclude, consumers by working all together and exploiting the digital connectivity, may minimize their overall cost and enhance their privacy.

4. Results

4.1. Problem Setup

The presented methodology is tested on a set of real-world data taken from the Irish power grid [33]. The dataset includes smart meter measurements of various residents. For testing our methodology, we select a set of smart meters and we assume that they belong to the same grid partition. In addition, we assume that the smart meters communicate and are able to apply the presented methodology to form their aggregated load signal. It should be noted, that the final output is the pattern that will be sent out to the utility as its final electricity order. Therefore, by "seizing" the final pattern, the utility as well as any third parties won't be able to make any inferences about the consumption activities of the individual consumers given that their activities will be masked by the morphed aggregated signal. In addition, the overall cost will be low, allowing consumer to "enjoy" cheaper consumption in a long-term horizon. Furthermore, ordering and scheduling of an aggregated load demand also allows consumers to perform last minute changes in their demand and exchange those changes without affecting the overall demand. For instance, someone might need to increase his load, while concurrently someone else might have to decrease it by the same amount—a situation that is promoted by the notion of virtual buffers described in [40,41].

In this work, in order to fully define the cost objective the utilization of a forecasted price signal is required. For simplicity, but without compromising the generalization, the forecasted signal is taken with the naïve method [42], which considers as forecasts the observed values of the same day a week ago. The forecasted as well as the real price signal, which the signals were taken from our previous work in [3], are depicted in Figure 4. Furthermore, we need to evaluate the morphing bounds of each consumer's anticipated load. In order to attain a diversity of the morphing values and exhibit a random behavior of the consumers, we implement a randomizer that provides random values of the upper and the lower morphing bounds. The randomizer provides a duet of values sampled from the following intervals:

- $[0.7 \times L_j^i, L_j^i]$ for the lower bound;
- $[L_j^i, 1.3 \times L_j^i]$ for the upper bound.

where L_j^i is the anticipated load of consumer j for the hour i.

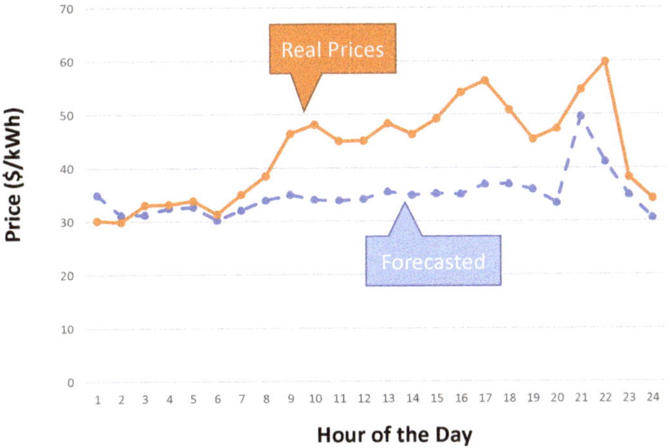

Figure 4. Real and forecasted price signal utilized for cost objective.

4.2. Tested Scenarios

In this section, the morphing methodology is applied on several scenarios and the respective results are recorded. In particular, we apply the presented methodology on grid partitions comprised six in the first scenario and of seven, 10 and 15 consumers in the rest of the scenarios. The performance of the methodology is assessed in two dimensions: The first dimension compares the overall cost provided by the morphing method to the cost before the morphing takes place. The second dimension measures the degree of privacy attained by the morphing methodology by comparing the correlation coefficients between the load patterns, i.e., between each of the initial individual patterns and the final morphed aggregated pattern. The results concerning the tested scenarios are presented in the subsections given below.

Scenario 1: Partition of Six Consumers

In this first scenario, we assume a small grid partition comprised of six consumers. Initially, each of the six consumers anticipates his/her day ahead load pattern; the respective load patterns for Consumers 1–6 are depicted in Figure 5. In the next step, the hourly morphing bounds of each of the consumers are obtained; the bounds pertained to the current scenarios are depicted in Figure 6 where both the upper and lower bounds are presented for all six consumers.

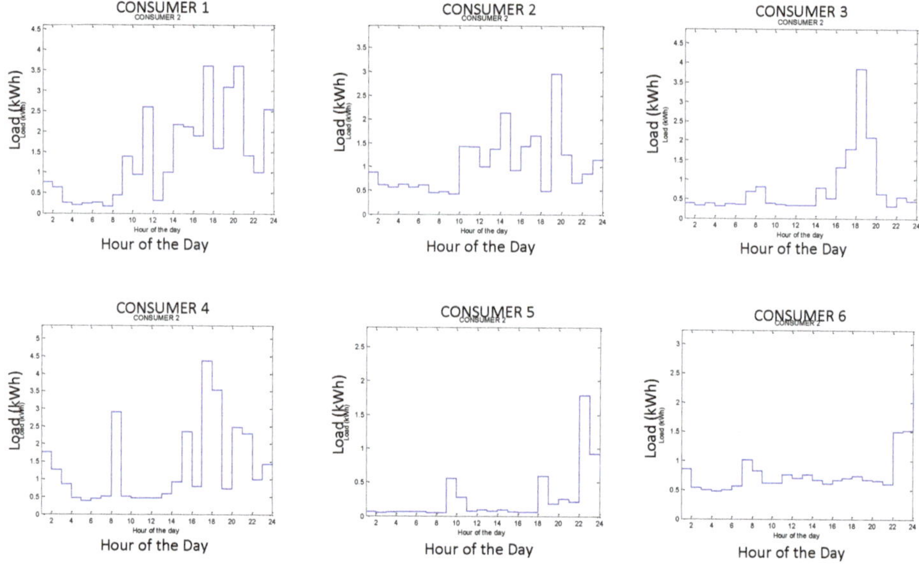

Figure 5. Anticipated load patterns of Consumers 1–6 for test Scenario 1.

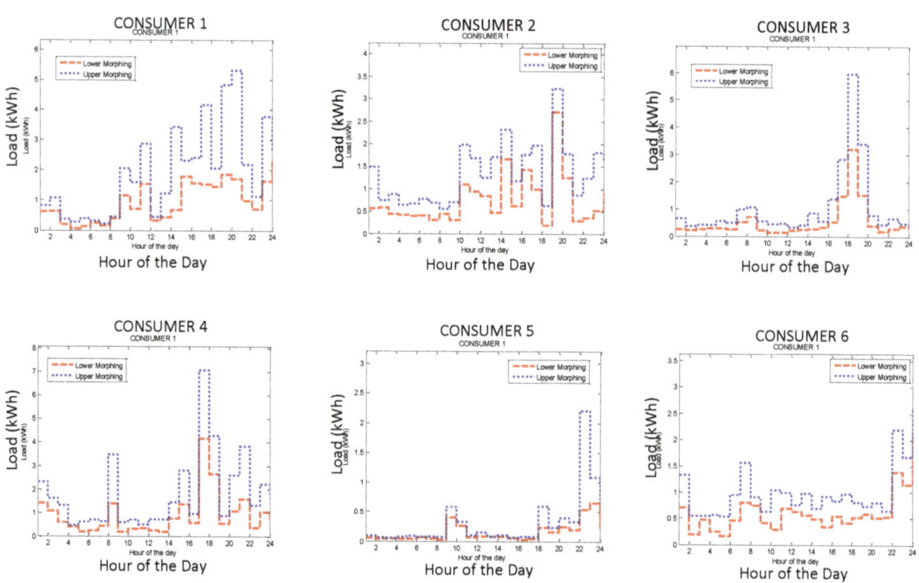

Figure 6. Anticipated load morphing bounds of Consumers 1–6 for test Scenario 1.

In the next step, the load patterns and their morphing bounds are aggregated to form the partition's load pattern and the respective morphing bounds. The latter ones, will comprise of the box constraints of the optimization problem, whose goal is to minimize the consumption cost as well as minimize the distance of the overall aggregated pattern to the its mean value. The aggregated pattern together with its morphing bounds are presented in Figure 7. In the following step, we compute the mean value of the aggregated pattern that is found to be equal to 5.69. Therefore, we create a constant pattern of the form:

$$M_i = 5.69 \quad i = 1, \ldots, 24 \tag{13}$$

i.e., a set of 24 values equal to 5.69. The pattern in Equation (13) is used to formulate the Mean Square Error (MSE) objective of the privacy measure.

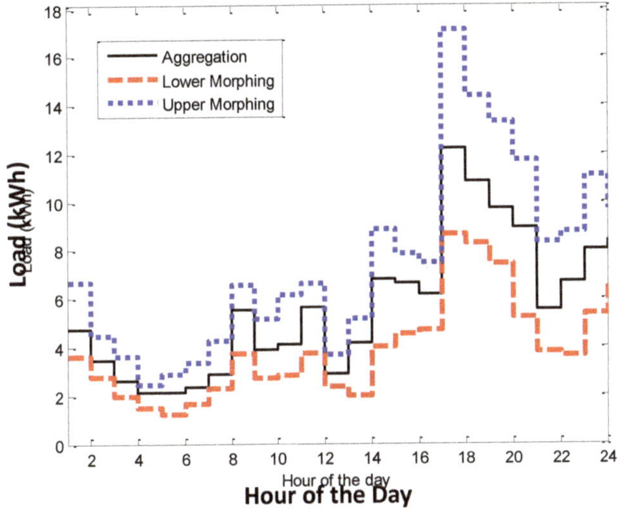

Figure 7. Aggregated load pattern of consumers 1–6 and the respective morphing upper and lower bounds.

At this point, the aggregated pattern and the morphing levels are utilized to set up the multi-objective problem, which takes the form of minimization problem. Firstly, the cost objective is defined by Equation (5), where the prices are given by Figure 4, while the load is the aggregated hourly load presented in Figure 7. Secondly, the privacy objective is given by Equation (9) where M = 0.59, and A_i^M is the aggregated hourly loads of Figure 7. Furthermore, the box constraints of the problem are taken by Equation (10) where the hourly lower and upper values are those presented in Figure 7.

Notably, the multi-objective formulation is fully defined and a solution is sought. To that end, the NSGA-II is applied to locating an optimal solution according to Pareto theory. The values of the genetic algorithm parameters, which are required to evolve the objectives, are taken as:

- Number of individuals: 30,
- Mutation probability: 0.01,
- Reproduction method: 15-point crossover,
- Max number of generations: 200,
- Selection of parents: Roulette method.

and the obtained solution is in the form of a vector whose length is equal to 24, i.e., one morphed load value for each hour of the day. The solution obtained in this specific scenario is provided in Table 1 (load values) and Figure 8 (plotted morphed pattern). We observe that the morphed aggregated pattern lies within the morphing bounds as determined in Figure 7 (as expected—thus validating our approach), while it differs from the initial aggregated pattern. Therefore, the algorithm has conducted a load morphing within the acceptable bounds that have been set by the grid partition consumers.

To fully assess the morphing methodology, we compute the costs and the privacy attained with the morphed pattern. In particular, we compute the following quantities:

- The non-morphed forecasted cost (NMFC), taken as the forecasted price multiplied by the aggregated load.
- The non-morphed real cost (NMRC), taken as the real price multiplied by the aggregated load.
- The morphed forecasted cost (MFC), given by the forecasted price multiplied by the morphed aggregated load.
- The morphed real cost (MRC), given by the real price multiplied by the morphed aggregated load.
- The correlation coefficient (CC) between each individual consumer pattern and the final morphed pattern denoting the degree of privacy achieved.

where we compute the costs of both forecasted and real price signals. The reason for doing that is to show that morphing the anticipated load not only reduces the anticipated cost but also the real cost.

Table 1. Load values in kWh taken as the solution of the multi-objective problem for Scenario 1.

Hour	1	2	3	4	5	6	7	8	9	10	11	12
Solution	4.87	4.10	3.48	1.53	1.80	2.77	4.11	5.60	2.89	3.04	3.83	2.54
Hour	**13**	**14**	**15**	**16**	**17**	**18**	**19**	**20**	**21**	**22**	**23**	**24**
Solution	3.11	4.54	4.74	5.94	9.21	8.63	8.09	5.72	5.66	4.13	5.46	6.61

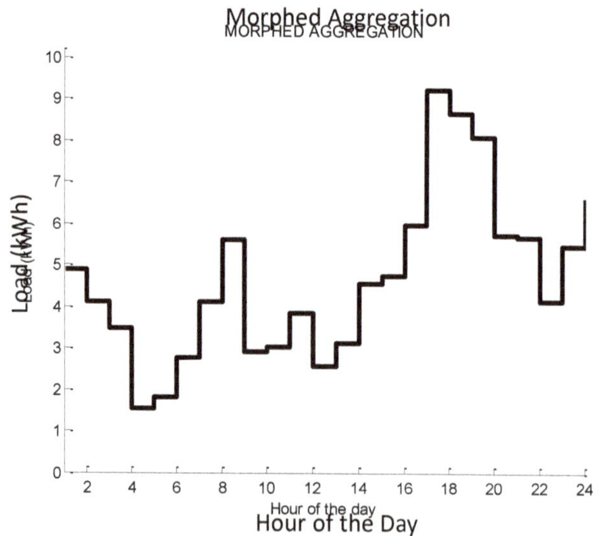

Figure 8. Plot of the final morphed aggregated load pattern computed by non-dominated sorting genetic algorithm-II (NSGA-II) (Values taken from Table 1).

Regarding the current scenario, the computed values are given in Table 2. By comparing the NMFC and the MFC, we clearly observe that the morphing methodology significantly reduced the anticipated expenses. In absolute numbers, the anticipated reduction is about $900 as we observe in Table 2. Furthermore, by comparing the real costs we observe that indeed a reduction in the electricity cost was attained. Notably, this reduction is about $1200, which far exceeds the anticipated reduction of the $900. Thus, we conclude that by morphing the aggregated load the consumers within the grid partition attained to reduce their actual aggregated expenses—emphatically, the real expense was much lower than the anticipated one.

Table 2. Cost and privacy values computed for Scenario 1. * Values expressed in US Dollars ($)

Quantity	NMFC	MFC	NMRC		MRC	
Value *	4837	3978	6200		5016	
Consumer	**#1**	**#2**	**#3**	**#4**	**#5**	**#6**
CC value	0.69	0.41	0.74	0.75	0.11	0.26

Regarding the privacy, we observe that the correlation coefficients between the initial consumers and the final morphed load has departed from the value of 1 (i.e., two patterns are fully correlated). In particular, we observe in Table 2 that the correlation values we obtain are lower than 0.76 implying that the morphed load shows less similarity to the consumers. The highest correlation is exhibited by consumers three and four whose correlation values are close to 0.75, followed by consumer one whose value is close to 0.7. The rest consumers exhibit very low correlation, quantified as values lower than 0.42, with consumer five exhibiting the lowest correlation that is equal to 0.11. Thus, we can state with much confidence that the morphed pattern differs from the individual ones, and thus masks the individual patterns. This difference, as quantified by correlation coefficient values, implies that an enhancement in each individual consumer privacy has been achieved, hence, imposing inference making of an individual's consuming activities highly challenging.

To sum up, in this first testing scenario, the six consumers via their collaboration, which is expressed in terms of load aggregation, attained to morph their initial anticipated load demand in such a way that the new patterns provides lower cost and higher privacy compared to the case of the individual non-morphed loads. This dual achievement was realized by evolving a multi-objective problem, which identified the optimal tradeoff between the cost and privacy objectives utilizing Pareto theory.

4.3. Further Results

In this section, the Pareto optimal morphing methodology is tested on a set of grid partitions comprised of various consumer numbers. As in the previous scenario, the data are also real-world data collected from smart meters deployed in Ireland [33]. In particular, our testing scenarios contain partitions comprised of seven (Scenario 2), 10 (Scenario 3) and 15 consumers (Scenario 4). Each test contains consumers of different load profiles, that have not been used in any of the other scenarios. Obtained results are recorded and given in Tables 3–5 respectively. It should be noted that in our previous work, i.e., [6,29], where we had focused only on consumption morphing for privacy issues, we had showed that the values of privacy measure of groups consisted of 15 consumers and above does not significantly get improved. In particular the studies showed that the more the consumers the higher the privacy is enhanced—for values of two to 10—and above 10 consumers, we observe a plateau in the privacy value curve [29], i.e., the values exhibited no significant increase. As a result, we have selected our testing cases to include up to 15 consumers—based on our previous work [29].

Observation of the obtained results in Tables 3–5 confirms the findings of previous section. The morphing of the aggregated pattern achieves lower overall consumption cost for the grid partitions. This is something that we observe in all three studied scenarios. This decrease in cost is observed both in forecasted cost and the actual (real) cost. Therefore, the presented methodology achieved lower cost in all tested scenarios independently of the number of participating consumers, and their profiles.

For comparison purposes, we have also implemented two alternative approaches: Both approaches define single objective problems. The first approach focuses on the privacy of the partition consumers, and aims at optimizing only the privacy objective. The second approach focuses on partition's electricity expenses, and aims at optimizing the cost objective. It should be noted that the optimization of the single objective problems is performed with a simple genetic algorithm.

Table 3. Cost ($) and privacy (correlation) measure values computed for Scenario 2 (Seven consumers).

	NMFC ($)	MFC ($)	NMRC ($)	MRC ($)	Average Correlation (between Morphed and Consumers)
Multi Objective (Our Method)	6248	5077	8104	6577	0.427
Single Objective: Consumer Cost	6248	5556	8104	7185	0.449
Single Objective: Consumer Privacy	6248	5908	8104	7603	0.451

Table 4. Cost ($) and privacy (correlation) measure values computed for Scenario 3 (10 consumers).

	NMFC ($)	MFC ($)	NMRC ($)	MRC ($)	Average Correlation (between Morphed and Consumers)
Multi Objective (Our Method)	6607	5635	8293	7039	0.269
Single Objective: Consumer Privacy	6607	6068	8293	7685	0.286
Single Objective: Consumer Cost	6607	6978	8293	8756	0.298

Table 5. Cost ($) and privacy (correlation) measure values computed for Scenario 4 (15 consumers).

	NMFC ($)	MFC ($)	NMRC ($)	MRC ($)	Average Correlation (between Morphed and Consumers)
Multi Objective (Our Method)	30,421	25,792	39,385	33,073	0.327
Single Objective: Consumer Cost	30,421	27,230	39,380	35,100	0.320
Single Objective: Consumer Privacy	30,420	29,401	39,380	37,730	0.330

By examining the computed correlation coefficients for the three scenarios, we also observe an enhancement of the consumers' privacy. In Scenario 2, all of the correlation values are lower than 0.63, providing an average value of 0.42. In Scenario 3, the highest correlation computed among consumers is equal to 0.8, which implicitly exhibits that all consumers have achieved a departure from the maximum correlation value of one. In addition, we observe that the average correlation is equal to 0.26 that implicitly exhibits that the final patterns show little resemblance with the initial patterns. Lastly, in scenario 4 we observe that the all consumers with the exception of consumer 12, whose cc value is 0.87, give correlation coefficients lower than 0.59. Notably, the average correlation in this scenario is found to be equal to 0.32, which also exhibits the significant departure from the max correlation of one. In sum, the morphing methodology contributes in enhancing the privacy of individual consumers; an enhancement expressed as a decrease of the correlation between the initial consumers' loads and the final morphed pattern. The average correlation per scenario is illustrative of the increased enhancement achieved by our methodology.

In comparison with the single objective cases, we also observe that our approach achieves lower electricity cost for the partition consumers, while it retains the average correlation between consumers and the final morphed pattern at the same level or slightly lower. With regard to privacy the single objective approach (i.e., in Tables 3–5 appears as "Single Objective: Consumer Privacy") provides an average correlation that is equal to 0.45, 0.29 and 0.33 for the scenarios of seven, 10 and 15 consumers, respectively. The aforementioned correlation values are pretty close to the ones obtained with our multi-objective approach demonstrating that utilization of multiple objectives does not deteriorate the privacy driven morphing. Moreover, we observe that by focusing on privacy, the partition cost significantly increases: This is observed in Tables 3–5 with respect to MFC and MRC values, where our approach provides significantly lower costs. With regard to the cost driven approach, which optimizes the cost objective (i.e., in Tables 3–5 appears as "Single Objective: Consumer Cost"), the obtained cost values are higher than those taken with our presented multi-objective approach for all three scenarios. Hence, those results validate the ability of our approach to optimize the consumption expenses of the grid partition. Overall, the comparison of the multi-objective approach with the single objective ones exhibits the ability of our presented approach to concurrently ensure low cost and high privacy in the partition grid. Thus, we conclude that formulation of a multi-objective optimization problem that concurrently handles cost and privacy is preferable to handling each of those objectives individually, given that it provides equal or higher performance.

5. Conclusions

In this paper, we have presented a new methodology for morphing the load pattern of a smart power grid partition. The partitions are assemblies of consumers that exploit the smart grid communications to collaborate and pursue common goals. Their goals entail minimization of their consumption expenses as well as enhancement of their consumption privacy.

The presented methodology achieves this set of goal by formulating a multi-objective problem. The problem comprises of two objectives, namely, the cost and the privacy objective. The first objective expresses the anticipated cost consumption for a day ahead of time, and the second measures the distance of the final load pattern to the initial consumers' anticipated load. The two objectives are minimized by an evolutionary algorithm and more specifically the NSGA-II that identifies a solution using the Pareto optimality theory. Furthermore, the presented work is compared to two single objective optimization approaches: The first approach handles only the cost objective, whereas the second approach handles the privacy objective. Comparison exhibited that our approach provides equal or better performance as compared to the single objective cases. It should be noted that for the privacy measure all three approaches provided very close values, proving that our approach enhances the degree of privacy as much as the privacy objective approach does. With regard to cost, the multi-objective morphing approach provided the lowest cost values in all tested cases. By combining the above observations, we conclude that the lumping of the two objectives in a single formulation did provide a better performance as compared to single objective problems. Therefore, we conclude that NSGA-II utilizing the Pareto theory attained to concurrently secure low cost and to enhance privacy.

However, the current study exhibits some limitations. More specifically, the optimization problem is based on the box constraints for each hour of the day. In this work, we assumed narrow intervals within the morphed values may lie. In practice, these intervals may be totally different than the ones assumed in this work, and depend on the characteristics of the grid and the consumers. Furthermore, we assume that the consumers are able to form a partition via direct communication links, and that the consumers trust each other (which is not always the case). Lastly, a limitation of our study has to do with the number of customers contained in the partition: We assume that the number of partition consumers remains constant at least for a whole day (since this is the anticipation horizon).

In the current work, the morphing methodology is tested on a set of real-world data taken from smart meters deployed in the Irish power grid. Testing entails four scenarios where grid partitions are comprised of six, seven, 10 and 15 consumers. Obtained results clearly demonstrate the effectiveness of the methodology in decreasing the partition's consumption costs (both anticipated and real cost), and enhancing the privacy of the individual consumers. Future work will focus on identifying and expanding the set of objectives that a grid partition may pursue.

Author Contributions: M.A. developed the concept and wrote the evolutionary computing code. N.G. created the datasets. Both authors contributed in writing the paper.

Funding: This research received no external funding.

Conflicts of Interest: The authors declare no conflict of interest.

References

1. Momoh, J. *Smart Grid: Fundamentals of Design and Analysis*; Wiley: New York, NY, USA, 2012; Chapter 1.
2. Tsoukalas, L.H.; Gao, R. From Smart Grids to an Energy Internet: Assumptions, Architectures, and Requirements. In Proceedings of the 3rd IEEE International Conference on Electric Utility Deregulation and Restructuring and Power Technologies, Nanjing, China, 6–9 April 2008; p. 98.
3. Alamaniotis, M.; Gatsis, N.; Tsoukalas, L.H. Virtual Budget: Integration of Load and Price Anticipation for Load Morphing in Price-Directed Energy Utilization. *Electr. Power Syst. Res.* **2018**, *158*, 284–296. [CrossRef]
4. Alamaniotis, M.; Tsoukalas, L.H.; Bourbakis, N. Virtual Cost Approach: Electricity Consumption Scheduling for Smart Grids/Cities in Price-Directed Electricity Markets. In Proceedings of the IISA 2014, Chania, Greece, 7–9 July 2014; pp. 38–43.
5. Wang, W.; Xu, Y.; Khanna, M. A survey on the communication architectures in smart grid. *Comput. Netw.* **2011**, *55*, 3604–3629. [CrossRef]
6. Alamaniotis, M.; Tsoukalas, L.H.; Bourbakis, N. Anticipatory driven nodal electricity load morphing in smart cities enhancing consumption privacy. In Proceedings of the IEEE Manchester PowerTech, Manchester, UK, 18–22 June 2017; p. 6.
7. Conejo, A.J.; Carrion, M.; Morales, L.M. *Decision Making Under Uncertainty in Electricity Markets*; Springer: London, UK, 2010.
8. Frederiks, E.R.; Stenner, K.; Hobman, E.V. Household energy use: Applying Behavioural Economics to understand Consumer Decision-Making and Behaviour. *Renew. Sustain. Energy Rev.* **2015**, *41*, 1385–1394. [CrossRef]
9. Anjos, M.F.; Gómez, J.A. Operations Research Approaches for Building Demand Response in a Smart Grid. In *Leading Developments from INFORMS Communities*; INFORMS: Catonsville, MD, USA, 2017; pp. 131–152.
10. Setlhaolo, D.; Xia, X.; Zhang, J. Optimal scheduling of household appliances for demand response. *Electr. Power Syst. Res.* **2014**, *116*, 24–28. [CrossRef]
11. Lujano-Rojas, J.M.; Monteiro, C.; Dufo-López, R.; Bernal-Agustín, J.L. Optimum residential load management strategy for real time pricing (RTP) demand response programs. *Energy policy* **2012**, *45*, 671–679. [CrossRef]
12. Oprea, S.V.; Bâra, A.; Diaconita, V. Sliding Time Window Electricity Consumption Optimization Algorithm for Communities in the Context of Big Data Processing. *IEEE Access* **2019**, *7*, 13050–13067. [CrossRef]
13. Oprea, S.V.; Bâra, A.; Ifrim, G. Flattening the electricity consumption peak and reducing the electricity payment for residential consumers in the context of smart grid by means of shifting optimization algorithm. *Comput. Ind. Eng.* **2018**, *122*, 125–139. [CrossRef]
14. Gatsis, N.; Giannakis, G.B. Residential demand response with interruptible tasks: Duality and algorithms. In Proceedings of the 2011 50th IEEE Conference on Decision and Control and European Control Conference, Orlando, FL, USA, 12–15 December 2011; pp. 1–6.
15. Oprea, S.V.; Bâra, A.; Reveiu, A. Informatics solution for energy efficiency improvement and consumption management of householders. *Energies* **2018**, *11*, 138. [CrossRef]
16. Baharlouei, Z.; Hashemi, M.; Narimani, H.; Mohsenian-Rad, H. Achieving optimality and fairness in autonomous demand response: Benchmarks and billing mechanisms. *IEEE Trans. Smart Grid* **2013**, *4*, 968–975. [CrossRef]
17. Safdarian, A.; Fotuhi-Firuzabad, M.; Lehtonen, M. A distributed algorithm for managing residential demand response in smart grids. *IEEE Trans. Ind. Inform.* **2014**, *10*, 2385–2393. [CrossRef]
18. Khodaei, A.; Shahidehpour, M.; Choi, J. Optimal hourly scheduling of community-aggregated electricity consumption. *J. Electr. Eng. Technol.* **2013**, *8*, 1251–1260. [CrossRef]

19. Verschae, R.; Kato, T.; Matsuyama, T. Energy management in prosumer communities: A coordinated approach. *Energies* **2016**, *9*, 562. [CrossRef]
20. Verschae, R.; Kawashima, H.; Kato, T.; Matsuyama, T. Coordinated energy management for inter-community imbalance minimization. *Renew. Energy* **2016**, *87*, 922–935. [CrossRef]
21. Shariatzadeh, F.; Mandal, P.; Srivastava, A.K. Demand response for sustainable energy systems: A review, application and implementation strategy. *Renew. Sustain. Energy Rev.* **2015**, *45*, 343–350. [CrossRef]
22. Chrysikou, V.; Alamaniotis, M.; Tsoukalas, L.H. A Review of Incentive based Demand Response Methods in Smart Electricity Grids. *Int. J. Monit. Surveill. Technol. Res.* **2015**, *3*, 62–73. [CrossRef]
23. Li, H.; Lin, X.; Yang, H.; Liang, X.; Lu, R.; Shen, X. EPPDR: An efficient privacy-preserving demand response scheme with adaptive key evolution in smart grid. *IEEE Trans. Parallel Distrib. Syst.* **2014**, *25*, 2053–2064. [CrossRef]
24. Gong, Y.; Cai, Y.; Guo, Y.; Fang, Y. A privacy-preserving scheme for incentive-based demand response in the smart grid. *IEEE Trans. Smart Grid* **2016**, *7*, 1304–1313. [CrossRef]
25. Wicker, S.; Thomas, R. A privacy-aware architecture for demand response systems. In Proceedings of the 44th Hawaii International Conference on System Sciences, Kauai, HI, USA, 4–7 January 2011; pp. 1–9.
26. Beligianni, F.; Alamaniotis, M.; Fevgas, A.; Tsompanopoulou, P.; Bozanis, P.; Tsoukalas, L.H. An internet of things architecture for preserving privacy of energy consumption. In Proceedings of the 10th Mediterranean Conference on Power Generation, Transmission, Distribution and Energy Conversion (Med Power 2016), Belgrade, Serbia, 6–9 November 2016; pp. 1–8.
27. Rahman, M.S.; Basu, A.; Kiyomoto, S.; Bhuiyan, M.A. Privacy-friendly secure bidding for smart grid demand-response. *Inf. Sci.* **2017**, *379*, 229–240. [CrossRef]
28. Li, D.; Aung, Z.; Williams, J.R.; Sanchez, A. No peeking: privacy-preserving demand response system in smart grids. *Int. J. Parallel Emerg. Distrib. Syst.* **2014**, *29*, 290–315. [CrossRef]
29. Alamaniotis, M.; Bourbakis, N.; Tsoukalas, L.H. Enhancing privacy of electricity consumption in smart cities through morphing of anticipated demand pattern utilizing self-elasticity and genetic algorithms. *Sustain. Cities Soc.* **2019**, *46*, 101426. [CrossRef]
30. Alamaniotis, M.; Tsoukalas, L.H.; Buckner, M. Privacy-driven electricity group demand response in smart cities using particle swarm optimization. In Proceedings of the 2016 IEEE 28th International Conference on Tools with Artificial Intelligence (ICTAI), San Jose, CA, USA, 6–8 November 2016; pp. 946–953.
31. Alamaniotis, M. Morphing to the Mean Approach of Anticipated Electricity Demand in Smart City Partitions Using Citizen Elasticities. In Proceedings of the 2018 IEEE International Smart Cities Conference (ISC2), Kansas City, MO, USA, 16–19 September 2018; pp. 1–7.
32. Alamaniotis, M.; Gatsis, N. Evolutionary Load Morphing in Smart Power System Partitions Ensuring Privacy and Minimizing Cost. In Proceedings of the 2018 Mediterranean Conference on Power Generation, Transmission, Distribution, and Energy Conversion (MEDPOWER 2018), Dubrovnik, Croatia, 12–15 November 2018; pp. 1–6.
33. Commission for Energy Regulation (CER). (2012). *CER Smart Metering Project—Electricity Customer Behaviour Trial, 2009–201 [dataset]*, 1st ed.; Irish Social Science Data Archive: Belfield, Irland, 2007; Available online: www.ucd.ie/issda/CER-electricity (accessed on 5 March 2019).
34. Deb, K.; Pratap, A.; Agarwal, S.; Meyarivan, T.A.M.T. A fast and elitist multiobjective genetic algorithm: NSGA-II. *IEEE Trans. Evol. Comput.* **2002**, *6*, 182–197. [CrossRef]
35. Ng, Y.K. The economic theory of clubs: Pareto optimality conditions. *Economica* **1973**, *40*, 291–298. [CrossRef]
36. Eiben, A.E.; Smith, J.E. *Introduction to Evolutionary Computing*; Springer: Berlin/Heidelberg, Germany, 2003.
37. Pareto, V. *Manuale di Economica Politica*; Societa Editrice Libraria: Milan, Italy, 1906; translated by Schwier, A.S., *Manual of Political Economy*; Schwier, A.S., Page, A.N., Eds.; Kelley: New York, NY, USA, 1971.
38. Marler, R.; Arora, J. Survey of multi-objective optimization methods for engineering. *Struct. Multidiscip. Optim.* **2004**, *26*, 369–395. [CrossRef]
39. Nasiakou, A.; Alamaniotis, M.; Tsoukalas, L.H.; Vavalis, M. Dynamic Data Driven Partitioning of Smart Grid Using Learning Methods. In *Handbook of Dynamic Data Driven Applications Systems*; Springer: Berlin/Heidelberg, Germany, 2018; pp. 505–526.
40. Alamaniotis, M.; Tsoukalas, L.H. Utilization of virtual buffer in local area grids for electricity storage in smart power systems. In Proceedings of the 2017 IEEE North American Power Symposium (NAPS), Morgantown, WV, USA, 17–19 September 2017; pp. 1–6.

41. Gao, R.; Tsoukalas, L.H. Implementing virtual buffer for electric power grids. In Proceedings of the International Conference on Computational Science, Beijing, China, 27–30 May 2007; Springer: Berlin/Heidelberg, Germany; pp. 1083–1089.
42. Alamaniotis, M.; Bargiotas, D.; Bourbakis, N.G.; Tsoukalas, L.H. Genetic optimal regression of relevance vector machines for electricity pricing signal forecasting in smart grids. *IEEE Trans. Smart Grid* **2015**, *6*, 2997–3005. [CrossRef]

Article

Modeling Phase Shifters in Power System Simulations Based on Reduced Networks †

Nuno Marinho [1,2,*], Yannick Phulpin [1], Adrien Atayi [1] and Martin Hennebel [2]

1 EDF R&D, 7 bd Gaspard Monge, F-91120 Palaiseau, France; yannick.phulpin@edf.fr (Y.P.); adrien.atayi@edf.fr (A.A.)
2 GeePs | Group of Electrical Engineering-Paris, 3, 11 Rue Joliot Curie, Plateau de Moulon, F-91192 Gif-sur-Yvette CEDEX, France; martin.hennebel@centralesupelec.fr
* Correspondence: nuno.marinho@edf.fr; Tel.: +33-1781-937-37
† This paper is an extension of the MEDPOWER 2018 conference paper.

Received: 29 March 2019; Accepted: 30 May 2019; Published: 6 June 2019

Abstract: Phase shifters are becoming widespread assets operated by transmission system operators to deal with congestions and contingencies using non-costly remedial actions. The setting of these controllable devices, which impacts power flows over large areas, may vary significantly according to the operational conditions. It is thus a key challenge to model phase shifters appropriately in power system simulation. In particular, accounting for the flexibility of phase shifters in reduced network models is a vibrant issue, as system stakeholders rely more and more on reduced models to perform studies supporting operational and investment decisions. Different approaches in the literature are proposed to model phase shifters in reduced network. Nevertheless, these approaches are based on the electrical parameters of the system which are not suitable for reduced network models. To address this problem, our paper proposes a methodology and assesses the impact of this contribution in terms of accuracy of the modelling on reduced network models. The approach was applied to a realistic case-study of the European transmission network that was clustered into a reduced network consisting of 54 buses and 82 branches. The reduction was performed using classical clustering methods and represented using a static power transfer distribution factor matrix. The simulations highlight that including an explicit phase shifter transformers representation in reduced models is of interest, when comparing with the representation using only a static power transfer distribution factor matrix.

Keywords: phase shifters; network reduction; power system operation; transmission network; power transfer distribution factor

1. Introduction

To deal with increasing uncertainty in system operation, Transmission System Operators (TSOs) rely more and more frequently on power flow control devices, such as Phase-Shifter Transformers (PSTs) or High-Voltage Direct Current (HVDC) links. These can be used as non-costly remedial actions to solve congestions by redirecting power flows, while avoiding redispatching or countertrading or in the longer-term the costly reinforcement of the network. It is therefore important to consider those assets appropriately when performing simulation of complex processes in large-scale power systems. As such, simulations are most frequently based on a simplified representation of the network (e.g., [1–3]), and modeling phase shifters in reduced networks has become a vibrant issue. Indeed, reduced networks generally aim to reflect the main steady-state features of the full system (e.g., [4]), but most studies perform static reductions based on a single operation point and therefore do not consider potential setting variations for the power flow control devices [5,6]. In [7], the authors considered multiple operation points in network reduction, but ended up with a reduced model based on static Power Transfer Distribution Factors (PTDF), independent of the setting of power flow control devices.

Different approaches are proposed in the literature to model a PST in load flow computations, either by modeling the PST as an impedance in series with an ideal transformer that varies with the tap angle [8], or by modeling its impact as an injected power, using a phase-shifter distribution factor (PSDF) matrix that establishes a relationship between the injected power and flow distribution through other network elements [9]. A limitation for these kinds of approaches is the need to know the electrical parameters of the system, namely lines' impedances or buses' angles. This can be a challenge when dealing with reduced network models, which can often be described using a PTDF matrix only.

To address this problem, this paper proposes a methodology to emulate and assess the impact of PSTs in reduced network models. In the reduced model, PSTs are represented as an extra variable that can be adjusted subject to the systems operating point, whereas the other network components are represented by a static PTDF matrix defined with the methodology presented in [7]. To assess the proposed methodology, multiple scenarios considering different operating conditions in both the reduced and complete models are simulated and branches' power flows are compared.

The approach was applied to a reduced network model based on the European power system using realistic parameters. The original European network data (available in [10]) with 2800 buses, 4000 branches and 4800 generation units were clustered to a 54-bus and 82-branch model and represented using a static PTDF matrix. The static PTDFs were defined based on 300 optimal power flows (OPFs) for various levels of net load to capture the seasonal characteristics of the European load profile.

The paper is organized as follows. Section 2 describes the clustering and connection modeling approach applied to obtain a reduced network model, details the methodology to define a PSDF and presents an illustrative example. Section 3 details the approach used to assess the impact of the proposed methodology. Section 4 presents a real scale test case and discusses the main results. Section 5 concludes.

2. Modeling Phase Shifters in a Reduced Model

2.1. Clustering

Historically, reduction techniques such as those in [11,12], relied on an ex-ante definition of buses to be maintained or aggregated, usually referred to as internal and external buses, respectively. These approaches were particularly successful for dynamic simulation models, where external buses were considered not to impact the internal buses and therefore would not be considered. Recent works apply clustering methodologies to aggregate buses with differences in the input information, i.e., topological [13] or economic data [14].

Following the observations in [15], where hierarchical clustering outperforms other clustering techniques for the aggregation of the European transmission network, hierarchical clustering was applied to aggregate the network buses in a determined number of zones. In this bottom-up algorithm, each network bus starts as an isolated zone, and at each step, the indicator A, as defined in Equation (1), is assessed and two electrically connected zones are aggregated until all the network buses belong to the same zone or the algorithm reaches the stopping criterion, which can either be a pre-determined number of zones or a threshold for the minimum distance between clusters.

The aggregation A is defined considering the local best scenario (minimum distance) at each stage, between the observations of Locational Marginal Prices (LMPs) for all the system's buses $n \in N$.

$$A = min(dist_{n,n'})\ \forall n \in N,\ \forall n' \in N \tag{1}$$

In [15], we suggested using the sum of the squared euclidean distances to determine the distance between pairs of observations for each scenario $s \in S$ as:

$$dist_{n,n'} = \sum_{s \in S} ||LMP_{n,s} - LMP_{n',s}||^2 \tag{2}$$

This algorithm presents a slow convergence for large datasets. On the other hand, given its bottom up approach, it guarantees electrical coherence of the aggregated buses.

2.2. Defining a PTDF Matrix for Static Network Assets

To model the transmission infrastructure connecting clusters, the most common input data are the nodal PTDFs of the complete network [5]. PTDFs describe the flow repartition through the entire system, caused by a potential change in power injection at a given bus. As detailed in [7], it is possible to determine a PTDF matrix (Ψ) and a set of loop flows (f^0) that reflect the flows between two clusters induced by the injected power within another cluster [16], while minimizing the difference between observed and estimated flows for each scenario s.

$$\min_{\{\Psi, f^0\}} (F_{l,s} - \bar{F}_{\bar{l},s})^2 \tag{3}$$

subject to

$$\forall s, \forall \bar{l} \quad \bar{F}_{\bar{l},s} = \sum_{\bar{n}} \Psi_{\bar{l},\bar{n}} \times P^{inj}_{\bar{n},s} + f^0_{\bar{l}} \tag{4}$$

$$\forall \bar{n}, \forall \bar{l} \quad |\Psi_{\bar{l},\bar{n}}| \leq 1 \tag{5}$$

Where:

- Variables:
  - Ψ is the PTDF matrix of dimension $\bar{L} \times \bar{N}$, with $\bar{L}$ and $\bar{N}$ being the reduced number of branches and buses, respectively;
  - $\bar{F}_{\bar{l},s}$ is the estimated flow in branch $\bar{l}$ for scenario s; and
  - $f^0_{\bar{l}}$ is the flow estimated error in branch $\bar{l}$ due to the aggregation of generation, denominated as loop flows.
- Parameters:
  - $F_{l,s}$ is the observed flow in branch l for scenario s; and
  - $P^{inj}_{\bar{n},s}$ is the power injected in bus $\bar{n}$ for scenario s.

2.3. Defining a PSDF Matrix for Phase Shifters

A PST introduces a difference in voltage angle between two nodes that can be modeled as a power injection through the branch where it is installed. This increase/decrease in the branch's flow affects the entire system with a redistribution over the other assets. In other words, considering that the tap position of a PST, corresponding to an angle δ, would reduce/increase the flow in a given branch l of a given system, to comply with Kirchoff's laws, this power variation ΔP_l should be distributed through the remaining branches of the system. This variation could be calculated using the coefficient from the system's PTDF matrix (Ψ) considering a as the injection bus and l the impacted branch. For example, $\Psi^a_{c,d}$ is the coefficient of the matrix representing the influence of the injection in bus a over the branch connecting buses c and d.

Considering a PST installed on the branch connecting buses a and d, as illustrated in Figure 1, the percentage of the new flow ($F^1_{(a,d)}$) that will be transferred to the branch connecting buses (c,d), $\omega_{(a,d)\rightarrow(c,d)}$ is determined as follows:

$$\omega_{(a,d)\rightarrow(c,d)} = \frac{\Delta F_{(c,d)}}{F^1_{(a,d)}} \tag{6}$$

The PST will induce an extra flow $\Delta P(\delta)$ that depends on the PST's angle installed on branch (a,d). Therefore, $F^1_{(a,d)}$ can be calculated as:

$$F^1_{(a,d)} = F^0_{(a,d)} + \Psi^a_{(a,d)} \times \Delta P(\delta) \tag{7}$$

which is the original flow on the branch ($F^0_{(a,d)}$) plus the power injected by the PST ($\Delta P(\delta)$) times the coefficient of the PTDF matrix for bus a in branch (a,d).

Considering that, in an extreme case, the injection by the PST equals the original flow on the branch, $F^0_{(a,d)} = \Delta P(\delta)$, Equation (7) becomes:

$$\Delta P(\delta) = \frac{F^1_{(a,d)}}{1 + \Psi^a_{(a,d)}} \tag{8}$$

The flow's increment on branch (c,d) due to the PST tap change is:

$$\Delta F_{(c,d)} = \Psi^a_{(c,d)} \times \Delta P(\delta) \tag{9}$$

which is the power injected by the PST ($\Delta P(\delta)$) times the coefficient of the PTDF matrix for bus a in branch (c,d).

Therefore, the impact of the PST can be calculated as:

$$\omega_{(a,d)\to(c,d)} = \frac{\Psi^a_{(c,d)}}{1 + \Psi^a_{(a,d)}} \tag{10}$$

For all the branches L of the system, the new flow F^1 can be calculated as:

$$F^1_l = F^0_l + \omega_l^{PST\ bus} \times \Delta P(\delta) \quad \forall l \in L \tag{11}$$

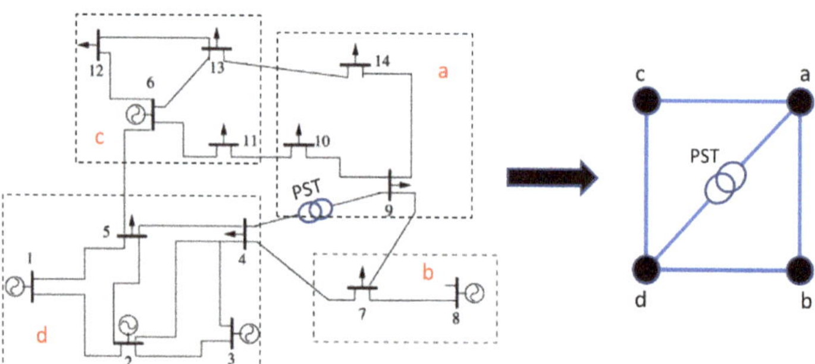

Figure 1. Illustrative example of the performance assessment for a reduction approach.

2.4. *Illustrative Example*

As illustrated in Figure 1, the Matpower's 14 bus test case [17] was modified to accommodate a PST between the buses 4 and 9. The original system buses were manually aggregated into four clusters and a static PTDF matrix was calculated to represent the power flows between them. The newly reduced system is composed of four buses connected by five branches, where the PST was installed in the branch connecting buses a and d. To assess the results, the five branches of the aggregated model were compared with the expected exchanges of the original system. For simplification purposes in notation, those aggregated exchanges were considered as five branches of the original system.

Considering that all branches have the same impedance, the PTDF matrix $\in \mathbb{R}^{L \times N}$ of the system is:

$$PTDF = \begin{bmatrix} 0.00 & -0.82 & -0.25 & -0.45 \\ 0.00 & -0.08 & -0.59 & -0.28 \\ 0.00 & -0.09 & -0.14 & -0.26 \\ 0.00 & 0.18 & -0.25 & -0.45 \\ 0.00 & -0.09 & 0.39 & -0.40 \end{bmatrix}$$

For a given set-point, the injected power $[P_{inj}]$ and flows on the system branches $[F]$ are:

$$P_{inj} = \begin{bmatrix} -53.40 & 8.50 & -34.30 & 79.20 \end{bmatrix}$$

$$F = \begin{bmatrix} -38.22 & -5.22 & -18.97 & -29.72 & -40.69 \end{bmatrix}$$

Applying Equation (10), a new matrix was defined that establishes the impact of the PST on every branch of the system:

$$PSDF = \begin{bmatrix} -0.36 & -0.22 & 1.00 & -0.36 & -0.24 \end{bmatrix}$$

Since in the reduced model the representation does not deal with the degrees, a choice was made to represent the MW value caused by the tap change. To do this, in the original system, a change of ± 1 deg in the PST installed in Branch 3 was made to find the associated power injection. Table 1 shows that $+1$ deg corresponds to an injection of 2.53 MW.

Table 1. Power flows for a PST with $\omega = \pm 1$ degree.

	−1 deg	**0 deg**	**+1 deg**
Branch 3	−17.40 MW	−14.87 MW	−12.34 MW

In this illustrative example, we intended to do a first assessment of the suitability of the PSDF matrix in representing PSTs in a reduced network model. Therefore, three load flows were run in the original system setting the PST tap position in 0, 2 and −2 degrees. The same was performed for the reduced system, this time with the linearization of the degrees into MW, as demonstrated in Table 1.

Table 2 shows the results for the same PST tap position in both the original and the reduced system. The same trends were followed in both models when the PST's tap positions were increased or decreased. As an example, the flow on Branch 1 decreased when the tap position was negative and the flow on Branch 3 increased in both models.

Table 2. Power flows for a PST with $\omega = \pm 2$ degree in both the original and reduced model.

	Original			**Reduced**		
	−2 deg	**0 deg**	**+2 deg**	**−5.06 MW**	**0 MW**	**+5.06 MW**
Branch 1	−29.22 MW	−32.23 MW	−35.25 MW	−35.14 MW	−38.22 MW	−41.30 MW
Branch 2	−4.88 MW	−6.58 MW	−8.29 MW	−3.31 MW	−5.22 MW	−7.14 MW
Branch 3	−20.30 MW	−15.58 MW	−10.87 MW	−24.03 MW	−18.97 MW	−13.91 MW
Branch 4	−20.72 MW	−23.73 MW	−26.75 MW	−26.64 MW	−29.72 MW	−32.80 MW
Branch 5	−39.18 MW	−40.88 MW	−42.59 MW	−38.64 MW	−40.69 MW	−42.74 MW

In addition, it can be observed that, in the reduced model, a linear relationship between the positive and negative tap setting was obtained. For example, in Branch 6, both the negative and positive tap setting implied a change of 3.08 MW, ensuring the linearity of the representation. It is important to note that the difference in the branches flows of both models when the PST was not activated was due to the loss of information in the reduction process and it was not related with the PSTs modeling.

3. Impact Assessment Metric

To assess the quality of the reduced model, a comparison of the estimated flows between the clusters $\bar{F}_{\bar{l},s}$ and the observed flows of the full model $F_{l,s}$ was made. To perform these simulations, two different sets of injected power were used:

1. P^{inj}_{zero} corresponds to the injected power issue of the full model simulation without any PST optimization.
2. P^{inj}_{optim} corresponds to the injected power issue of the full model simulation with the optimization of the PST. During this calculation, all PSTs of network were considered as an optimization variable that could vary within ±30 degrees.

In addition, three different reduced model representations were considered:

1. Ψ_{PST} was calculated using the methodology presented in [7], and having as input P^{inj}_{optim}. This model does not explicitly include any variable to represent the PSTs. In other words, this is a static PTDF matrix that was built using the injected power from the full model simulation with the optimization of the PST.
2. Ψ_{static} was calculated using the methodology presented in [7], and having as input P^{inj}_{zero}.
3. $\Psi_{static} + PSDF$, where, besides the Ψ_{static} matrix, a $PSDF$ matrix was also calculated using the methodology described in Section 2.3.

With these cases, it was intended to highlight the effects of the PSTs on different reduced network models.

A first assessment was performed using a PTDF matrix following the methodology described in [7] (Ψ_{static}). As this does not explicitly models the PST, one can assess the error of the proposed methodology when PSTs are optimized. To do that, the different scenarios were simulated, one where PSTs were optimized (P^{inj}_{optim}) and other they were not considered (P^{inj}_{zero}).

Once the accuracy of the model when PSTs are neglected is known, the accuracy of explicitly representing PSTs in the reduced model is assessed, as suggested in Section 2.3. To do that, a PTDF matrix Ψ_{static} and a PSDF matrix that describes the impact of PSTs were used to describe the system.

The differences between the estimated $\bar{F}_{\bar{l},s}$ and observed $F_{l,s}$ flows were compared using the Root Mean Square Error (RMSE):

$$RMSE = \frac{1}{S_E}\sum_{s=0}^{S_E}\sqrt{\frac{\sum_{l,\bar{l}=0}^{L}(F_{l,s}-\bar{F}_{\bar{l},s})^2}{L}} \tag{12}$$

where S_E is the set of evaluation scenarios and L the total number of interconnectors.

In addition, to have an overview of the methodology performance over the extreme scenarios of the proposed cases, a Value at Risk (VaR) index was calculated to assess the risk of extreme under performance for a reduced set of scenarios.

4. Case-Study

The proposed approach was applied to a realistic large scale power system to assess the impact of PST of the branches' power flows.

4.1. Case Description

4.1.1. Network Data

The focus was on a realistic European network model [10] with 2842 400 kV and 225 kV buses, 1820 generators and 3739 branches, representing 12 countries, namely France (FR), Belgium (BE),

Netherlands (NL), Germany (DE), Austria (AT), Switzerland (CH), Italy (IT), Slovenia (SL), Poland (PL), Czech Republic (CZ), Slovakia (SK) and (West) Denmark (DK). Five other countries were modeled with a unique bus (i.e., United Kingdom, Greece, Sweden, Norway and (East) Denmark). The original network was updated to account for the inclusion of new branches reported on the 2014 TYNDP report [6].

4.1.2. Load Data

Historic 2013 hourly load demand profiles for each country were used, which were downloaded from the ENTSO-E transparency website [18] and then allocated to each bus based on the original load distribution of the model. Total load values and highest peak can be found in Table 3.

Table 3. Total load and highest peak for each country in 2013.

Country	Total Load (TWh)	Highest Peak (GW)
DE	590.80	91.80
AT	64.40	10.00
BE	86.60	13.40
FR	445.90	82.80
NL	103.70	16.40
PL	143.10	23.20
CZ	66.60	10.10
CH	61.90	9.80
IT	290.50	52.50
SL	12.30	1.90
SK	24.60	3.80
DK	30.50	5.60

4.1.3. Generators Data

Using the commercial database *PLATTS* [19], containing all the technical information required, generators data were updated to a 2013 scenario.

In Table 4, the installed capacities of the non-dispatchable generation per country, namely Photovoltaic (PV), Wind Power Onshore (WP On), Wind Power Offshore (WP Off), Combined Heat and Power (CHP) and Hydro Power Run-of-River (HP RoR), are presented. Annual profile samples for each country were collected from [18]. The installed capacity of the dispatchable generation per country is presented in Table 5.

Table 4. Capacity of non-dispatchable generation in *GW*.

Country	PV	WP On	WP Off	CHP	HP RoR
DE	32.00	31.00	0.10	17.50	2.30
AT	0.40	1.30	0.00	1.70	4.30
BE	2.70	1.00	0.40	1.30	0.10
FR	3.60	8.00	0.10	3.20	6.10
NL	0.40	2.30	0.20	4.20	0.00
PL	0.00	2.30	0.00	5.90	0.20
CZ	2.00	0.30	0.00	5.20	0.20
CH	0.20	0.10	0.00	0.20	1.60
IT	15.20	6.70	0.00	4.10	2.30
SL	0.20	0.00	0.00	0.20	0.30
SK	0.50	0.00	0.00	1.00	1.40
DK	0.30	2.60	0.40	0.70	0.00

Table 5. Capacity of dispatchable generation in *GW*.

Country	Nuclear	Coal	Fuel	Gas	Lignite	Reservoir
DE	12.00	18.80	2.50	18.10	20.50	7.80
AT	0.80	0.00	0.40	3.30	4.30	8.00
BE	3.90	0.90	1.20	5.70	0.00	1.30
FR	63.00	7.00	6.90	4.00	0.00	18.50
NL	0.50	3.90	13.20	0.00	0.00	0.00
PL	0.00	15.30	0.40	0.30	9.70	1.70
CZ	3.80	0.80	0.00	0.10	4.90	1.70
CH	3.20	0.00	0.10	0.10	0.00	11.70
IT	0.00	7.20	11.60	47.00	0.20	15.20
SL	0.70	0.50	0.00	0.00	0.80	0.80
SK	1.90	0.40	0.00	0.50	0.50	1.00
DK	1.70	0.00	0.10	1.00	0.00	0.00

4.2. Reduced Model

The hierarchical clustering, as defined in Section 2.1, was applied to aggregate the network, using 300 different LMP scenarios.

The LMP scenarios were calculated using 300 operating points corresponding to different levels of net load for the overall system to capture the seasonal characteristics of the European load profile. Those 300 periods were uniformly picked from the annual European load curve, whose total energy values per country can be found in Table 3. During this calculation, all PSTs of network were considered as an optimization variable that could vary between −30 and +30 degrees.

The goal of the network reduction methodology was to keep the information regarding network congestions, with a much smaller representation. For sake of simplicity, the aggregation was performed separately for each existing bidding zone and only for the buses at its interior, avoiding clusters that would share different bidding zones. Despite that, the methodology allowed aggregating buses regardless of the bidding zone definition.

In addition, since the goal of this approach was to study the impact of PST, a supplementary constraint was added r to avoid the aggregation of areas connected by a PST. Therefore, the algorithm would stop when all observations at the interior of a zone were grouped except for those connected by a PST.

This resulted in an equivalent model with 54 buses and 82 branches. The connectivity between clusters were defined to match those of the complete model. The same input data were used to define the reduced PTDF matrix, as detailed in Section 2.2.

4.3. Impact of PST Modeling

Section 2.4 demonstrates the suitability of the PSDF matrix to represent PST in reduced network models. The goal was to assess the accuracy of the representation in a large-scale power system with realistic values. Therefore, the same approach was used: the full European transmission network was clustered into a set of reduced buses and represented by a static PTDF matrix, as stated above, and then the same tap optimization was applied in both the full network model (with an explicit representation of the PST) and in the reduced model (using the PSDF matrix). The results in terms of branches' power flows were compared to determine the best PST representation.

Given the high complexity of analyzing the entire power system, for simplification purposes, we focused on analyzing a single PST located at the border between Germany and the Netherlands. As detailed in Section 3, three different cases ertr used to assess the proposed methodology:

1. Ψ_{PST} is a static PTDF matrix calculated having as input P^{inj}_{optim} and does not explicitly include any variable to represent the PSTs.
2. Ψ_{static} is a static PTDF matrix calculated having as input P^{inj}_{zero} and does not explicitly include any variable to represent the PSTs.

3. $\Psi_{static} + PSDF$ includes the previous PTDF matrix Ψ_{static} and a $PSDF$ matrix.

The generation plan was firstly obtained by running an OPF with the full model and was then used to perform a load flow with the reduced model. The resulting flows of both the OPF and load flow were compared to assess the accuracy of the modeling.

Table 6 presents the error of the flows obtained with the reduced network model for Ψ_{static} and Ψ_{PST} PTDF matrix modeling. The error was calculated using Equation (12) with the evaluation scenarios S_E.

Table 6. Error values for the power flows comparison between the full and reduced model, without an explicit PST modeling.

	P^{inj}_{zero}	P^{inj}_{optim}
Ψ_{static} (MW)	143.30	200.60
Ψ_{PST} (MW)	204.70	138.50

Table 6 demonstrates the impact of the optimization of the PST in the reduced model. The Ψ_{static} matrix performed well when the input scenarios did not consider the optimization of the PST, but the error tended to increase when the P^{inj}_{optim} was used. On the other hand, the Ψ_{PST} matrix could reduce the error for the case where the PST was optimized P^{inj}_{optim}, with a RMSE of only 138.5 MW per branch per scenario, but the error rapidly increased when P^{inj}_{zero} was applied.

The results in Table 6 show that the static PTDF matrix representation performed well under the scenarios from which it was built. When the "non PST optimized" injected powers (P^{inj}_{zero}) were applied to the "PST optimized" PTDF matrix (Ψ_{PST}), the results are less accurate than when applied to the "non PST optimized" PTDF matrix (Ψ_{static}) representation and vice versa.

The same trend can be observed in Table 7, which presents the VaR of 5% for the flows calculated using the reduced model. It can be remarked that the values of VaR were similar for both P^{inj}_{zero} and P^{inj}_{optim} when using the Ψ_{static}, but more significant values arose when applying P^{inj}_{zero} to the Ψ_{PST} matrix.

Table 7. Value at risk of 5% for the flows comparison between the full and reduced model, without an explicit PST modeling.

	P^{inj}_{zero}	P^{inj}_{optim}
Ψ_{static} (MW)	273.10	279.30
Ψ_{PST} (MW)	307.20	248.90

Figures 2 and 3 show the error distribution for the Ψ_{static} and Ψ_{PST} PTDF matrix, respectively, when applying P^{inj}_{zero} and P^{inj}_{optim}. Following the same trend observed in Tables 6 and 8, the results demonstrate that, when the operational scenarios with no PSTs modeled (P^{inj}_{zero}) were applied to the non PST optimized PTDF matrix (Ψ_{static}), the errors were lower than when the opposite occurred. With these two figures, the over fitting process that occurred in the optimization process defined in Section 2.2 became clear, therefore, stressing the need for a more general representation of PSTs in the reduced models.

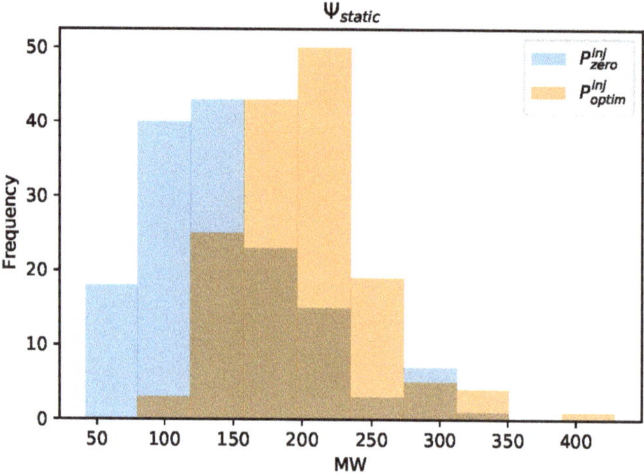

Figure 2. Error distribution for the static PTDF matrix representation when applying the P^{inj}_{zero} and P^{inj}_{optim} datasets.

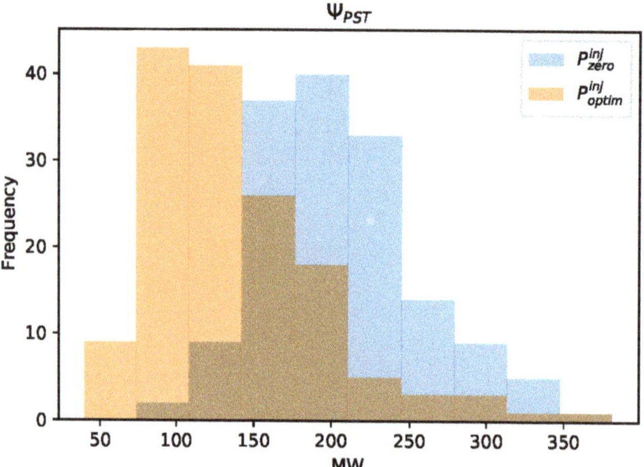

Figure 3. Error distribution for the PST optimized PTDF matrix representation when applying the P^{inj}_{zero} and P^{inj}_{optim} datasets.

Table 8. Performance indexes for the reduced model, with an explicit PST modeling.

$\Psi_{static} + PSDF$	P^{inj}_{zero}	P^{inj}_{optim}
RMSE (MW)	143.30	194.20
VaR (MW)	273.10	275.30

Finally, the reduced model including an explicit modeling of the PST ($\Psi_{static} + PSDF$) was tested. For the injected power set where PSTs were optimized (P^{inj}_{optim}), the PST of the reduced model mimicked its behavior; in other words, an injected power was multiplied by the $PSDF$ matrix. As presented in

Section 2.4, a relationship between degrees of the PST and injected power was established and the same power was injected by the PST in the reduced model.

Table 8 shows the error results for the case where an explicit modeling of the PST was done (Ψ_{static} + $PSDF$). As can be observed, for the case where the PST was not optimized (P^{inj}_{zero}), the error was the same as presented in Table 6, as the injected power of the PST of the reduced model was set to zero.

When considering the case where the PST was optimized (P^{inj}_{optim}), the error showed a slight reduction issue to the explicit PST modeling. In addition, when looking into the VaR, it was observed that, with the explicit modeling, the VaR tended to be similar independently of the considered case.

Figure 4 shows the error distribution for the representation using a PTDF + PSDF matrix. It was observed that the error obtained with the P^{inj}_{zero} was similar to the one presented in Figure 2. For the P^{inj}_{optim}, the error was superior to the one obtained with the Ψ_{PST} PTDF matrix (as shown in Figure 3) but inferior to the ones obtained with the Ψ_{static} PTDF matrix.

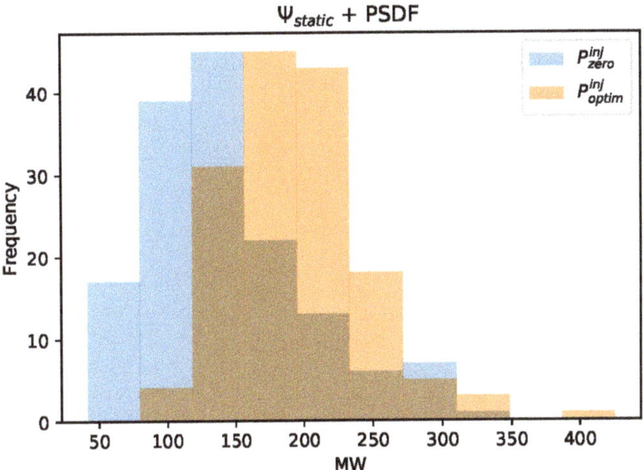

Figure 4. Error distribution for the explicit PST representation when applying the P^{inj}_{zero} and P^{inj}_{optim} datasets.

Comparing both the results obtained with and without the explicit PST modeling, it was observed that relying only on the optimization of the PTDF matrix using a set of data where the PSTs were optimized tended to underperform the case where the PST were not optimized and vice versa. When adding the explicit modeling of the PSTs, the results for the case using P^{inj}_{optim} were not as accurate as the ones obtained with the Ψ_{PST} matrix, but were more accurate for the case where P^{inj}_{zero} was applied.

Overall, the proposed model with an explicit modeling of the PSTs lost some accuracy for a specific case P^{inj}_{optim}, but compensated for the opposite case P^{inj}_{zero}, as highlighted by the VaR values presented in Table 8.

5. Conclusions

This paper presents a simplified methodology to model PSTs in reduced network models and assesses its effects in transmission networks. The proposed methodology is applied to a real test case allowing to identify challenges that are not easily addressed with simplified test systems. A performance assessment index is proposed that allows assessing the pertinence of representation in a reduced model. Three cases were studied, including the one where no PST representation existed.

Preliminary results tend to show promise for modeling PSTs in reduced network models. The simulations showed that explicitly modeling phase shifters along with the PTDF matrix could increase the level of accuracy of the reduced model compared to the approach based only on the definition of the PTDF matrix. In our view, this highlights the interest of the proposed methodology for the development of reduced static model of a large scale power network, such as the ones performed in [6]. The further deployment of these devices over the network triggers more and more difficulties to a static representation of the power system over its different operation conditions.

It is important to stress that, given the specificities of power systems, results are conditioned by the choice of the simulated PST. A PST in a more central position or next to critical bottlenecks can have a different impact on the system flows and production plan, in the same way as a more decentralized PST can cause the inverse. In addition, it is important to remember that all results were obtained using the DC approximation and should therefore be carefully interpreted as the results of a full AC analysis might differ.

A key direction for further work is to expose the proposed methodology to a larger set of operating conditions, and equipment specificities. It would also be of interest to assess the suitability of such methodology to model HVDC lines and assess its impact on reduced network models.

Author Contributions: Conceptualization, N.M., Y.P., A.A. and M.H.; Methodology, N.M., Y.P., A.A. and M.H.; Validation, Y.P. and M.H.; Writing—original draft preparation, N.M.; Writing—review and editing, N.M. and Y.P.; Supervision, M.H.; Project administration, A.A.

Funding: This research received no external funding.

Conflicts of Interest: The authors declare no conflict of interest.

Abbreviations

The following abbreviations are used in this manuscript:

CHP	Combined Heat and Power
HP RoR	Hydro Power Run-of-River
HVDC	High-Voltage Direct Current
LMP	Locational Marginal Prices
OPF	Optimal Power Flow
PTDF	Phase-Shifter Distribution Factor
PST	Phase-Shifter Transformer
PTDF	Power Transfer Distribution Factors
PV	Photovoltaic
RMSE	Root Mean Square Error
TSO	Transmission System Operator
VaR	Value at Risk
WP Off	Wind Power Offshore
WP On	Wind Power Onshore

References

1. e–HIGHWAY 2050. *D 2.2—European Cluster Model of the Pan-European Transmission Grid*; Technical Report; European Commission: Brussels, Belgium, 2014.
2. Lumbreras, S.; Ramos, A. The new challenges to transmission expansion planning. Survey of recent practice and literature review. *Electr. Power Syst. Res.* **2016**, *134*, 19–29. [CrossRef]
3. Shayesteh, E.; Hobbs, B.F.; Amelin, M. Scenario reduction, network aggregation, and DC linearisation: Which simplifications matter most in operations and planning optimisation? *IET Gener. Transm. Distrib.* **2016**, *10*, 2748–2755. [CrossRef]
4. Wang, L.; Klein, M.; Yirga, S.; Kundur, P. Dynamic reduction of large power systems for stability studies. *IEEE Trans. Power Syst.* **1997**, *12*, 889–895. [CrossRef]
5. Shi, D.; Tylavsky, D. A Novel Bus-Aggregation-Based Structure-Preserving Power System Equivalent. *IEEE Trans. Power Syst.* **2015**, *30*, 1977–1986. [CrossRef]
6. ENTSO-E. *10-Year Network Development Plan*; ENTSO-E: Bruxelles, Belgium, 2014.

7. Marinho, N.; Phulpin, Y.; Folliot, D.; Hennebel, M. Network Reduction Based on Multiple Scenarios. In Proceedings of the 2017 IEEE Manchester PowerTech, Manchester, UK, 18–22 June 2017; pp. 1–6.
8. ENTSOE–E. *Phase Shift Transformers Modelling*; Technical Report; ENTSO-E: Bruxelles, Belgium, 2014.
9. Hertem, D.V. The Use of Power Flow Controlling Devices in the Liberalized Market. Ph.D. Thesis, KU Leuven, Leuven, Belgium, 2009.
10. ENTSO-E Common Grid Model. Available online: https://www.entsoe.eu/major-projects/commoninformation-model-cim/cim-for-grid-models-exchange/standards/Pages/default.aspx (accessed on 1 July 2018).
11. Ward, J.B. Equivalent Circuits for Power-Flow Studies. *Trans. Am. Inst. Electr. Eng.* **1949**, *68*, 373–382. [CrossRef]
12. Dimo, P. *Nodal Analysis of Power Systems*; Abacus Bks, Editura Academiei: Bucharest, Romania, 1975.
13. Lumbreras, S.; Ramos, A.; Olmos, L.; Echavarren, F.; Banez-Chicharro, F.; Rivier, M.; Panciatici, P.; Maeght, J.; Pache, C. Network partition based on critical branches for large-scale transmission expansion planning. In Proceedings of the 2015 IEEE Eindhoven PowerTech, Eindhoven, The Netherlands, 29 June–2 July 2015; pp. 1–6. [CrossRef]
14. Burstedde, B. From nodal to zonal pricing: A bottom-up approach to the second-best. In Proceedings of the 2012 9th International Conference on the European Energy Market (EEM), Florence, Italy, 10–12 May 2012; pp. 1–8. [CrossRef]
15. Marinho, N.; Phulpin, Y.; Folliot, D.; Hennebel, M. Redispatch index for assessing bidding zone delineation. *IET Gener. Transm. Distrib.* **2017**, *11*, 4248–4255. [CrossRef]
16. THEMA. *Loop Flows–Final Advice*; Technical Report; European Comission: Brussels, Belgium, 2013.
17. Zimmerman, R.D.; Murillo-Sanchez, C.E.; Thomas, R.J. MATPOWER: Steady-State Operations, Planning, and Analysis Tools for Power Systems Research and Education. *IEEE Trans. Power Syst.* **2011**, *26*, 12–19. [CrossRef]
18. ENTSO-E Transparency Platform. Available online: https://transparency.entsoe.eu/ (accessed on 26 May 2018).
19. PLATTS World Electric Power Plants Database. Available online: https://www.spglobal.com/platts/en/products-services/electric-power/world-electric-power-plants-database (accessed on 26 May 2018).

Article

Preventing Internal Congestion in an Integrated European Balancing Activation Optimization

Martin Håberg [1,2,*], Hanna Bood [2] and Gerard Doorman [2]

1 Dept. of Electric Power Engineering, Norwegian University of Science and Technology, 7491 Trondheim, Norway
2 Statnett SF, N-0423 Oslo, Norway; hanna.bood@statnett.no (H.B.); gerard.doorman@statnett.no (G.D.)
* Correspondence: martin.haberg@ntnu.no; Tel.: +47-934-26-391

Received: 7 January 2019; Accepted: 29 January 2019; Published: 3 February 2019

Abstract: New common platforms for optimization of balancing energy activation will facilitate cross-border exchange and integrate the fragmented European balancing markets. Having a zonal market structure, these platforms will optimize balancing actions as if intra-zonal transmission constraints did not exists, leaving it to each Transmission System Operator (TSO) to manage internal congestion caused by balancing energy activations. This paper describes a new method to pre-filter balancing bids likely to cause internal congestion due to their location. Furthermore, the complementary concept of exchange domains has been developed to prevent congested and infeasible balancing situations. A numerical example illustrates both the effectiveness and limitations of each method.

Keywords: balancing market design; congestion management; optimization methods; power system modeling

---

1. Introduction

As European balancing markets are being integrated, common methodologies and systems are being developed to optimize activation and exchange of balancing energy across borders. Under the current target model, each TSO will submit their balancing energy needs for the upcoming imbalance settlement period, as well as a list of available bids within their own area, to a common European platform. The common platform aims to identify the most efficient set of bid activations to cover the imbalances in all areas.

Congestion in the transmission grid incurs the risk of overloads, and must be managed to avoid endangering operational security. Congestion between different market areas can be prevented through cross-zonal capacity constraints in the platform optimization, but intra-zonal—or internal—congestion may occur as a result of bid locations and initial power flows in the network. The enormous size of the interconnected European system, the limited available time in the operational phase, and the preference of zonal market coupling bodes that European balancing platforms will not include the highly detailed network models necessary to represent internal bottlenecks. Rather, the common balancing platform will select the balancing actions, but the task of managing internal congestion is left to each TSO.

This paper presents two methodologies for TSOs to prevent internal congestion caused by an integrated European balancing activation optimization. Firstly, a method for *bid filtering* is described. Based on extensive power flow analyses across a variety of potential situations, the method aims to detect potentially harmful bid activations and flag the corresponding bids as unavailable. Secondly, the paper introduces a new concept of *exchange domains*, complementing the bid filtering by ensuring feasibility and enabling more bids to be made available to the common platform.

Section 2 summarizes the development of an integrated European balancing market, and highlights earlier contributions to managing internal congestion from balancing activations. The two congestion management mechanisms are described in the subsequent sections. Section 3 explains pre-filtering balancing energy bids, introducing a new, multi-dimensional approach to assess which bids to make available to a common European balancing platform. The concept of exchange domains is introduced and explained in Section 4. Following a numerical example illustrating both methods in Section 5, the paper concludes in Section 6 with a discussion on the merits and viability of each of the concepts for a future integrated European balancing market.

2. Background

Over the last few years, the European Network of Transmission System Operators for Electricity (ENTSO-E) have developed new network codes, rules and regulations for European power markets. In particular, the Guideline on Electricity Balancing [1] specifies the target model for an integrated European balancing market, including standardization of balancing energy products across countries to facilitate exchange. Aiming to increase efficiency in resource utilization, balancing energy bids located in different areas will be collected into *common merit order lists* (CMOLs), from which an Activation Optimization Function (AOF) will select bids for activation to cover the imbalances of all TSOs, taking into account possibilities for netting of imbalances and available cross-zonal transmission capacity between areas.

Several European TSOs are collaborating in balancing pilot projects to develop and implement common activation and exchange optimization platforms for the different reserve products. Notably, the *TERRE* (Trans European Replacement Reserves Exchange) project establishes a platform for cross-border exchange of balancing energy from replacement reserves (RR) [2], while the *MARI* (Manually Activated Reserves Initiative) [3] and *PICASSO* (Platform for the International Coordination of Automated Frequency Restoration and Stable System Operation) [4] projects develop European platforms for the exchange of balancing energy from frequency restoration reserves with manual activation (mFRR) and automatic activation (aFRR), respectively. The available time between optimization and activation will be limited, restricting the possibilities for redispatch. All of these projects suggest preventing balancing actions from causing internal congestion by letting each TSO mark bids as unavailable if their activation could endanger system security, in accordance with Art. 29.14 in [5].

Congestion management is a central aspect of a zonal electricity market design, and is necessary when the price structure does not reflect the impact of grid congestion, as compared to locational marginal prices (LMPs). For zonal markets, Linnemann et al. [6] identified three main mechanisms to manage congestion: grid expansion, market splitting, and redispatch. Only the latter can be applied in an operational timeframe, and it is used in several European systems to manage internal congestion.

Several factors can impact power flows, and thereby potentially also network congestion during the balancing stage. Contingencies and power imbalances from intermittent generation are inherently stochastic. Interestingly, the balancing market itself may also have a strong effect, depending on the imbalance pricing mechanism. In a study of the German balancing market, Chaves-Ávila et al. [7] argued that using a single area-wide imbalance price signal may be adverse and misleading in the presence of internal congestion, worsening the local imbalance in part of the system, with the potential result of further congesting the network. At the same time, there may be very limited time and flexibility to effectively manage congestions through redispatch during the balancing stage.

Another crucial factor is the impact of bid activations in the balancing energy market. Some systems allow portfolio-based bids, meaning the exact locations of balancing energy injections are often unknown. For the German power system, Sprey et al. [8] concluded that the effect on congestion from reserve activation is unforeseeable, and uses simulation to assess the impact. In the Norwegian system, on the other hand, the location of each balancing bid is largely known. This allows the effect on network flows from bid activation to be predicted using power flow analyses. This is used in [9] to develop an algorithm that evaluates whether balancing bids must be skipped in the merit order to satisfy requests for balancing energy exchange from different neighboring zones. This can support decisions on which bids to make available to a European balancing platform, yet the proposed algorithm is one-dimensional, only considering balancing energy requests from a single neighboring zone at a time, i.e., no combinations or transit requests.

Finally, concerns on intra-zonal network constraints are not limited to European market designs. In the US, traditional reserve requirements have partitioned the grid into deterministic reserve zones, mainly based on ad-hoc rules [10]. Disregarding the intra-zonal constraints, and thereby the grid location of reserves procured within a zone, incurs the risk of ineffective means to handle intra-zonal congestion [11]. Moreover, since all reserves within a zone are assumed to have equal shift factors on critical lines, the true deliverability of the procured reserves will be imprecise. Acknowledging that different contingencies render different reserves undeliverable, Lyon et al. [12] demonstrated a locational reserve disqualification method to ensure adequate volumes and locations of operating reserves to cope with a range of distinct scenarios.

3. Bid Filtering

A zonal market platform will optimize balancing actions as if intra-zonal transmission constraints did not exist, in some cases leading to the activation plans that are infeasible due to internal congestion. In an attempt to prevent infeasible activation plans, the *MARI* platform [3] plans to allow TSOs to mark individual bids as unavailable if their activation would lead to internal congestion. Thus, each TSO needs to assess—in advance of the platform clearing—whether activating a bid would lead to congestion or not. The impact of balancing bid activations on internal congestion depends not only on the bid location, but also on the location of the request, as well as the current (or predicted) flow in the intra-zonal network. Guntermann et al. [9] showed on a realistic dataset how bid activations can often cause congestion when requested from one or more of the neighboring zones, while causing no congestion if requested from other zones.

The bid filtering methodology proposed in this paper aims to determine the availability of balancing energy bids within a given bidding zone, with each neighboring area considered to be represented by a single external node (cf. Figure 1). It extends the work in [9] by considering combinations of balancing requests from multiple neighboring areas. While this is more realistic, it also increases complexity. Moreover, since these requests cannot be accurately predicted, the proposed method needs to consider a range of combined balancing energy requests from the immediate neighboring zones of a given zone. Each request combination is denoted as an *exchange scenario*, and the method evaluates for each of the scenarios whether avoiding internal congestion requires deviating from merit order activation.

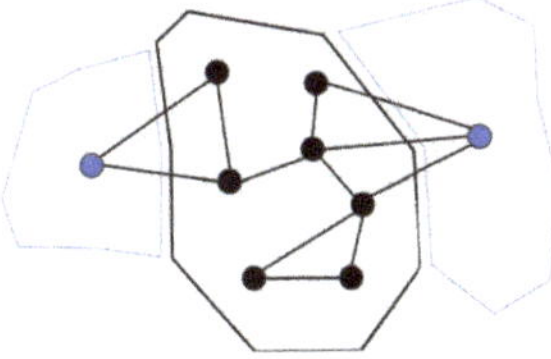

Figure 1. Example single-area system consisting of internal nodes and external nodes representing neighboring areas.

To evaluate each exchange scenario, a local balancing activation optimization problem is solved on a detailed network model. For a given zone a, the objective function in Equation (1) minimizes the activation cost given by the bid price C_b and activation volume y_b of each bid b available in the local bid list $\mathcal{B}_a$. The energy balance constraint in Equation (2) requires for each internal or external node $i \in \mathcal{I}_a$ that the net imbalance E_i is covered either by flow f_l or bid activation from bids located at node i. The adjacency parameter $A_{il} \in \{-1, 0, 1\}$ ensures adequate connections between each node i and lines $l \in \mathcal{L}_a$. Depending on the directions and volumes of imbalance volumes, upward or downward activations can be disallowed (e.g., by making sets $\mathcal{B}_i^{\uparrow}$ and $\mathcal{B}_i^{\downarrow}$ empty for all i) to prevent simultaneous counter-activations within a zone. The flow constraint in Equation (3) translates the balancing energy injection in each node to balancing energy flows f_l on each line l, through the power transfer distribution factor (PTDF) Φ_{il}, while Equation (4) limits the activation volume of each bid to its capacity $\overline{Y}_b$. Balancing energy flows are limited by the remaining available capacities $\underline{F}_l$ and $\overline{F}_l$ on each line in Equation (5).

$$\min_{f,y} \sum_{b \in \mathcal{B}_a} C_b y_b \tag{1}$$

$$\text{s.t.} \sum_{b \in \mathcal{B}_i^{\uparrow}} y_b - \sum_{b \in \mathcal{B}_i^{\downarrow}} y_b - \sum_{l \in \mathcal{L}_a} A_{il} f_l + E_i = 0, \qquad i \in \mathcal{I}_a \tag{2}$$

$$f_l - \sum_{i \in \mathcal{I}_a} \Phi_{il} \Big(\sum_{b \in \mathcal{B}_i^{\uparrow}} y_b - \sum_{b \in \mathcal{B}_i^{\downarrow}} y_b + E_i \Big) = 0, \qquad l \in \mathcal{L}_a \tag{3}$$

$$0 \le y_b \le \overline{Y}_b, \quad b \in \mathcal{B}_a \tag{4}$$

$$\underline{F}_l \le f_l \le \overline{F}_l, \quad l \in \mathcal{L}_a \tag{5}$$

The congesting bids can be identified by considering the resulting nodal balancing prices in each evaluated scenario. The Lagrangian multiplier λ_i on each energy balance constraint in Equation (2) provides a locational marginal price on balancing energy in each node i for the minimum-cost feasible balancing dispatch. If no internal transmission constraints are binding, this dispatch will follow the merit order. If, on the other hand, congestion prevents bids from being used in the merit order, this will be visible through shadow price differences between different nodes. The locational balancing energy price on the external nodes indicate the marginal costs of exporting one more unit of balancing energy to the corresponding neighboring zone. If there is unused upward capacity with a bid price lower than these marginal exchange costs, this bid is congesting the system. The same is true with opposite price differences for congested downward resources. In these cases, it is clear that the bid price does not reflect the full cost of activation, and following the merit order would have been infeasible due to transmission constraints. In short, the bids not fully utilized although priced within the marginal cross-zonal price are the ones that would cause congestion in the particular scenario if activated in the merit order.

Bids causing congestion only when activated under special circumstances provide a dilemma. Filtering such bids from the list reduces the reserve capacity and increases balancing costs in situations where they could have been used, after all. Not filtering them would lead to congestion and distorted price signals in some cases. The detection of congested bids in individual scenarios does not provide a final answer as to which bids to make available to the platform. However, it provides insight on the degree to which each bid causes congestion when it is activated, in some cases suggesting that the bid should be filtered from the list.

The computational burden of the bid filtering process depends, among other things, on the number of exchange scenarios to be evaluated. The structure of Equations (1)–(5) is linear and largely similar to a DC OPF problem, and requires minimal computational effort. Even with time requirements in the near-operational phase, this structure should allow using detailed network models and a substantial number of exchange scenarios. The scenario selection used for the numerical example in Section 5 constitutes a trivial approach, using a matrix of equidistant exchange volumes. The size of these intervals will affect both the number of scenarios and the accuracy of the results. However, with the set of merit-order feasible scenarios forming a convex region, sensitivity analysis would require only a subset of these scenarios to be evaluated for each bid list configuration. Moreover, the method does not require exhaustive enumeration of all possible bid list configurations, but iteratively removes one bid at a time and evaluates whether more scenarios become merit-order feasible.

4. Exchange Domains

Given the bid locations and initial flow in the network, there are exchange scenarios for which internal congestion cannot be avoided. Should such a scenario materialize, the congestion must be managed through an urgent redispatch to avoid disconnections. Another approach is to attempt to prevent infeasible scenarios from materializing. Rather than filtering bids, a more adequate solution to this end would be to identify and disallow unfavorable combinations of cross-border flows.

The key idea of exchange domains is to add additional constraints in the platform optimization on balancing energy exchange volumes to neighboring areas. This can be used to eliminate the possibility of exchange requests that are found to be infeasible in the exchange scenario evaluation. Furthermore, scenarios where deviation from the merit order is necessary can also be discarded in this manner, thereby enabling many bids to be made available without causing the platform to give incorrect price signals. Constraints describing exchange domains would need to be submitted to the platform together with the list of available bids.

The selection of an exchange domain for a given area can be based on the same exchange scenario analyses as for bid filtering, and needs to take into account the final list of available bids. A robust approach is to select a domain such that all scenarios are included for which merit order activation of the available bids is feasible. Since lists of upward and downward bids are used exclusively in their direction of activation, upward and downward domains must be considered separately as well. A possible step-by-step method is summarized below.

4.1. Determine List of Available Bids

The exchange domain will be tailored towards a filtered list of bids. In principle, any bid filtering method can be applied before this step.

4.2. Evaluate Exchange Scenarios

Precalculating the local balancing dispatch for different combinations of balancing energy exchange provides a discrete approximation of the feasibility region. This enables identifying the borderline of feasibility, and also where the available bids can be used in the merit order.

4.3. Convex Hull Transformation

Each evaluated exchange scenario can be represented as a point, with coordinates given by the balancing exchange volumes to neighboring zones in the particular scenario. If the set S contains all points representing exchange scenarios evaluated as merit-order feasible for the filtered bid list, then the convex hull (Convex hulls are efficiently calculated from a finite set of points using the *Quickhull* algorithm [13], even for higher dimensions.) $Conv(S)$ is the smallest convex polytope containing all these points.

4.4. Define Linear Constraints

Each facet in the convex hull corresponds to a supporting hyperplane defining a half-space, and all the points in S are enclosed by the intersection of these half-spaces. The inequalities describing each half-space directly comprises a finite set of linear constraints, efficiently describing the feasible region of exchange situations, or exchange domain.

5. Numerical Example

Based on the work in [14], this numerical example highlights the important steps in the bid filtering method and the relation to exchange domains. A test system based on the IEEE 30-bus network is used, with two of the nodes (7 and 30) assumed to represent neighboring zones (cf. Figure 2). The flow in the transmission network is already initialized by an economic dispatch, with no lines being initially congested. Six generators at different nodes serve 156 MW of local load, in addition exporting 23 MW and 11 MW on exchange nodes 7 and 30, respectively.

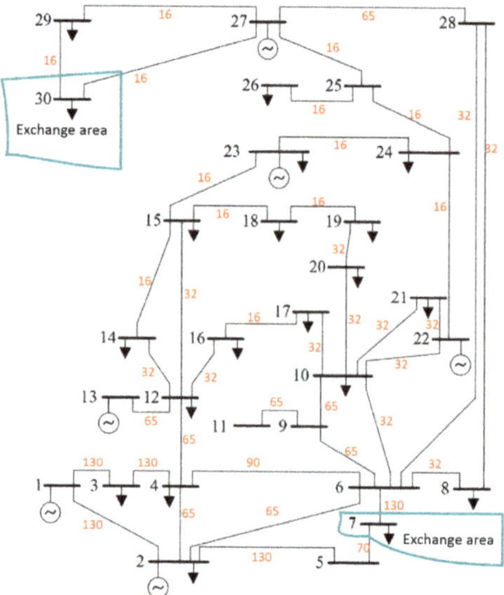

Figure 2. Single line diagram of the test system used, indicating bus numbers (black) and rated line capacities in MW (orange).

5.1. *Bid Filtering*

This example focuses on determining the availability of six upward bids. Downward bids could have been evaluated using broadly the same methodology [14], and have been omitted here for brevity. All bids have a capacity of 10 MW, and carry names *Bid #1–Bid #6* according to their position in the merit order, given by bid prices.

The selection of which exchange scenarios to evaluate should in principle cover all possible balancing exchange outcomes from the European balancing platform, i.e., limited by cross-zonal capacities to neighboring zones in both directions. Without sufficient bid capacity to cover many of the resulting rather extreme scenarios, this example considers a reduced set of exchange scenarios, given by the net injections $E_7 \in [-90, 10]$ and $E_{30} \in [-20, 40]$ at the exchange nodes, with 10 MW steps between scenarios (cf. Figure 3a).

Upon evaluating these scenarios using the optimization in Equations (1)–(5), only a subset of them can be balanced given the initial flow in the network and the bid list at hand. The cells with numerical values in Figure 3b represent exchange scenarios for which there exists a feasible balancing dispatch. Some of these scenarios are congested (shown in red), and require deviating from the merit order, i.e., one or more bids must be (at least partially) skipped to avoid overloading the network.

Comparing shadow prices on balancing energy exchange with bid prices in the different scenarios reveals that one of the bids, *Bid #4*, located on bus 22, causes congestion in several of the congested scenarios. These scenarios are marked as red in Figure 4a. Filtering this bid from the list and re-evaluating the exchange scenarios shows the scenarios previously congested by Bid #4 are now merit-order feasible (cf. Figure 4b).

In this example, the three red scenarios for $E_{30} = 20$ cannot be made merit-order feasible by filtering any of the bids. Moreover, there is no efficient way of filtering bids to avoid most infeasible scenarios. For example, preventing the combination $(E_7, E_{30}) = (-40, 30)$ requires making *all upward bids* unavailable. This scenario requires only 10 MW of activated balancing energy, so reserve capacity is not an issue here, but the transit flow is.

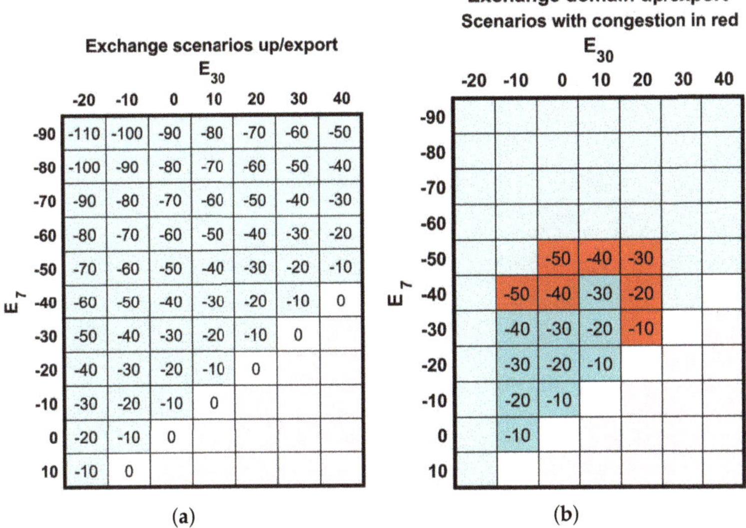

Figure 3. Net balancing activation volumes in the considered exchange scenarios: (**a**) All exchange scenarios; (**b**) Feasible scenarios.

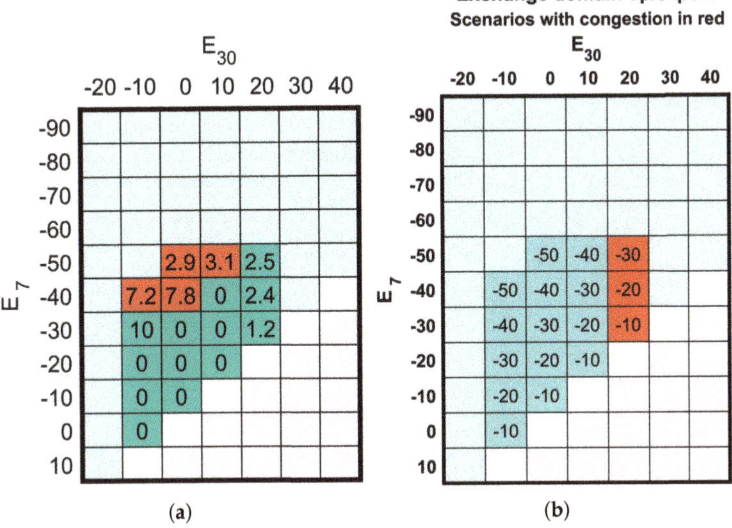

Figure 4. (**a**) Activation volume of Bid #4, causing congestion in red scenarios; (**b**) Feasible exchange scenarios after filtering Bid #4.

5.2. Exchange Domains

Following the steps in Section 4, an exchange domain can be specified to prevent infeasible and congested exchange flow combinations. Here, exchange domains are calculated for the full upward bid list, and a list where Bid #4 is filtered.

5.2.1. All Six Upward Bids Made Available

The linear constraints corresponding to the convex hull is found using the Quickhull algorithm [13] with a list of coordinates corresponding to feasible exchange scenarios. Assuming only merit-order feasible scenarios should be included, these scenarios correspond to the green cells with numeric values in Figure 3b. The linear constraints enclosing the desired domain will be

$$
\begin{aligned}
E_{30} &\geq -10 \\
E_{30} &\leq 10 \\
E_7 + 0.5E_{30} &\geq -35
\end{aligned}
$$

Although computed by Quickhull, these constraints can easily be manually verified in this example, since the vertices of the convex hull can be identified directly from Figure 3b as $(-10, 0)$, $(-10, -30)$, $(10, -40)$ and $(10, -20)$.

5.2.2. Bid #4 Made Unavailable

After withholding Bid #4, a larger set of exchange scenarios become merit-order feasible. (cf. Figure 4b). The scenarios in the red cells are still congested. Using Quickhull on the list of merit-order feasible scenarios eventually provides the hyperplanes constraining the feasible region (cf. Figure 5).

$$
\begin{aligned}
E_{30} &\geq -10 && \text{(I)} \\
E_{30} &\leq 10 && \text{(II)} \\
E_7 &\geq -50 && \text{(III)} \\
E_7 + E_{30} &\geq -50 && \text{(IV)}
\end{aligned}
$$

Compared to the case with all bids available, the convex hull has different vertices: $(-10, 0)$, $(-10, -40)$, $(0, -50)$, $(10, -50)$ and $(10, -20)$. One more linear constraint is needed, but the resulting exchange domain is larger.

Exchange domain up/export
Scenarios with congestion in red

E_7 \ E_{30}	-20	-10	0	10	20	30	40
-90							
-80							
-70							
-60							
-50			-50	-40	-30		
-40		-50	-40	-30	-20		
-30		-40	-30	-20	-10		
-20		-30	-20	-10			
-10		-20	-10				
0		-10					
10							

Figure 5. Exchange domain defined by linear constraints after filtering Bid #4.

The same procedure also applies for areas with more neighboring areas, albeit with more variables due to higher dimensionality of the exchange scenarios. Including these constraints to the common platform optimization enables making the balancing bids available without the risk of activation or transit flow leading to internal congestion.

6. Discussion

The impact of balancing energy activation on internal congestion depends strongly on the location of the imbalance. With the possibility of the balancing energy recipient being several different combinations of neighboring zones, bid locations cannot be seen as singular predictors for internal congestion due to balancing actions. Cross-border balancing exchange flows, including transit flows, appear to have similar, or higher importance.

For the balancing platforms being developed, filtering congested bids is currently the preferred mechanism for avoiding internal congestion. The method for bid filtering proposed in this paper takes the uncertainty in exchange flows into account by considering a large number of discrete exchange scenarios. For each scenario, a balancing dispatch followed by a nodal price analysis detects which bids would cause congestion with the given exchange flows. While bids causing congestion are obvious candidates for being filtered from the common merit order list, the picture is rarely black and white; bids can lead to congestion under some scenarios while being perfectly safe to use in many others.

The evaluation of an exchange scenario for a given bid list distinguishes between three outcomes: infeasible, congested, or merit-order feasible. While the latter indicates that the available bids can safely be activated for the exchange flows at hand, congested scenarios require skipping bids and deviating from the merit order. Infeasible scenarios simply cannot be balanced with any combination of bids from the list, and redispatch would be critical to avoid overloading the transmission network should such a scenario materialize. Whereas making specific bids unavailable can make congested scenarios merit-order feasible, bid filtering is ineffective in preventing infeasible platform outcomes.

Exchange domains provide restrictions on cross-zonal balancing energy flow combinations. These cannot make more exchange scenarios feasible, but effectively prevent infeasible or congested situations from occurring. In this regard, the concept is complementary to bid filtering. An exchange domain must be tailored to the list of available bids, and the domains calculated for specific lists of bids can also help determine which bids to filter. More importantly, the ability to discard congested scenarios without filtering bids also enables making more bids available to the platform.

An exchange domain is described by a set of linear inequalities, and these would act as additional constraints in the platform optimization. The impact on the computational burden from these additional constraints would be negligible. The concept is newly developed and thus has not been proposed as a candidate method in the design drafts of any of the European balancing platforms being implemented. Nevertheless, the importance of cross-border balancing flows on internal congestion infers that including such a mechanism should be considered in the future.

Author Contributions: The paper was partly based on the MSc thesis of H.B., with individual contributions including: conceptualization, M.H.; methodology, M.H. and H.B.; software, H.B.; validation, H.B., M.H. and G.D.; formal analysis, H.B.; investigation, H.B.; resources, H.B., M.H.; data curation, H.B.; writing—original draft preparation, M.H.; writing—review and editing, M.H.; visualization, H.B.; supervision, M.H. and G.D.; project administration, G.D.; and funding acquisition, G.D.

Funding: This research was funded by the Norwegian Research Council grant number 248291.

Conflicts of Interest: The authors declare no conflict of interest. The funders had no role in the design of the study; in the collection, analyses, or interpretation of data; in the writing of the manuscript, or in the decision to publish the results.

Abbreviations

The following abbreviations are used in this manuscript:

aFRR	Automatically activated Frequency Restoration Reserves
AOF	Activation Optimization Function
CMOL	Common Merit Order List
ENTSO-E	European Network of Transmission System Operators for Electricity
LMP	Locational Marginal Price
mFRR	Manually activated Frequency Restoration Reserves
MARI	Manually Activated Reserves Initiative
PICASSO	Platform for the International Coordination of Automated Frequency Restoration and Stable System Operation
PTDF	Power Transfer Distribution Factor
RR	Replacement Reserves
TERRE	Trans European Replacement Reserves Exchange
TSO	Transmission System Operator

References

1. European Commission. *Guideline on Electricity Balancing*; European Commission: Brussels, Belgium, 2018.
2. ADMIE; National Grid; REE; REN; RTE; Terna; SwissGrid. Public Consultation Document for the Design of the TERRE (Trans European Replacement Reserves Exchange) Project Solution. 2016. Available online: https://consultations.entsoe.eu/markets/public-consultation-document-for-the-design-of-the/results/terre-public-consultation-assessment.pdf (accessed on 3 February 2019).
3. MARI. First Consultation—Call For Input. 2017. Available online: https://consultations.entsoe.eu/markets/mari-first-consultation-call-for-input/supporting_documents/20171121_%20MARI%20First%20Consultation_Final.pdf (accessed on 3 February 2019).
4. 50 Hertz; Amprion; APG; Elia; RTE; TenneT; Transnet BW. Consultation on the Design of the pLatform for Automatic Frequency Restoration Reserve (aFRR) of PICASSO Region. 2017. Available online: https://docstore.entsoe.eu/Documents/Network%20codes%20documents/Implementation/picasso/PICASSO-Consultation_document.pdf (accessed on 3 February 2019).
5. ENTSO-E. Guideline on Electricity Balancing (May 2017 Version). 2017. Available online: https://electricity.network-codes.eu/network_codes/eb/ (accessed on 3 February 2019).
6. Linnemann, C.; Echternacht, D.; Breuer, C.; Moser, A. Modeling optimal redispatch for the European Transmission grid. In Proceedings of the 2011 IEEE Trondheim PowerTech, Trondheim, Norway, 19–23 June 2011; number chapter IV, pp. 1–8. [CrossRef]
7. Chaves-Ávila, J.P.; van der Veen, R.A.C.; Hakvoort, R.A. The interplay between imbalance pricing mechanisms and network congestions—Analysis of the German electricity market. *Util. Policy* **2014**, *28*, 52–61. [CrossRef]
8. Sprey, J.D.; Drees, T.; vom Stein, D.; Moser, A. Impact of balancing energy on network congestions. In Proceedings of the 2015 12th International Conference on the European Energy Market (EEM), Lisbon, Portugal, 19–22 May 2015; pp. 1–6.
9. Guntermann, C.; Gunderson, N.W.; Lindeberg, E.; Håberg, M. Detecting unavailable Balancing Energy Bids due to Risk of Internal Congestions. In Proceedings of the 18th International Conference on Environment and Electrical Engineering (EEEIC), Palermo, Italy, 12–15 June 2018.
10. Wang, F.; Hedman, K.W. Reserve zone determination based on statistical clustering methods. In Proceedings of the 2012 North American Power Symposium (NAPS), Champaign, IL, USA, 9–11 September 2012; pp. 1–6. [CrossRef]
11. Lyon, J.D.; Hedman, K.W.; Zhang, M. Reserve Requirements to Efficiently Manage Intra-Zonal Congestion. *IEEE Trans. Power Syst.* **2014**, *29*, 251–258. [CrossRef]
12. Lyon, J.D.; Zhang, M.; Hedman, K.W. Locational reserve disqualification for distinct scenarios. *IEEE Trans. Power Syst.* **2015**, *30*, 357–364. [CrossRef]

13. Barber, C.B.; Dobkin, D.P.; Huhdanpaa, H. The quickhull algorithm for convex hulls. *ACM Trans. Math. Softw.* **1996**, *22*, 469–483. [CrossRef]
14. Bood, H. Exchange Domains for Cross-border Activation of Balancing Bids—Preventing Internal Congestion in an Integrated EUropean Balancing Activation Optimization. Master's Thesis, Norwegian University of Science and Technology, Trondheim, Norway, 2018.

Article

Digital Forensic Analysis of Industrial Control Systems Using Sandboxing: A Case of WAMPAC Applications in the Power Systems †

Asif Iqbal *, Farhan Mahmood and Mathias Ekstedt

School of Electrical Engineering and Computer Science, KTH Royal Institute of Technology, Stockholm SE-100 44, Sweden

* Correspondence: asif.iqbal@ee.kth.se

† This paper is an extension of the paper Iqbal, A., Mahmood, F., & Ekstedt, M. (2018, November), An Experimental Forensic Testbed: Attack-based Digital Forensic Analysis of WAMPAC Applications. Presented at the 11th Mediterranean Conference on Power Generation, Transmission, Distribution, and Energy Conversion (MEDPOWER 2018), Dubrovnik, Croatia, 12–15 November 2018.

Received: 11 May 2019; Accepted: 3 July 2019; Published: 6 July 2019

Abstract: In today's connected world, there is a tendency of connectivity even in the sectors which conventionally have been not so connected in the past, such as power systems substations. Substations have seen considerable digitalization of the grid hence, providing much more available insights than before. This has all been possible due to connectivity, digitalization and automation of the power grids. Interestingly, this also means that anybody can access such critical infrastructures from a remote location and gone are the days of physical barriers. The power of connectivity and control makes it a much more challenging task to protect critical industrial control systems. This capability comes at a price, in this case, increasing the risk of potential cyber threats to substations. With all such potential risks, it is important that they can be traced back and attributed to any potential threats to their roots. It is extremely important for a forensic investigation to get credible evidence of any cyber-attack as required by the Daubert standard. Hence, to be able to identify and capture digital artifacts as a result of different attacks, in this paper, the authors have implemented and improvised a forensic testbed by implementing a sandboxing technique in the context of real time-hardware-in-the-loop setup. Newer experiments have been added by emulating the cyber-attacks on WAMPAC applications, and collecting and analyzing captured artifacts. Further, using sandboxing for the first time in such a setup has proven helpful.

Keywords: forensic investigations; forensic evidence substation; wide area monitoring protection and control; phasor measurement units (PMUs); industrial control systems; sandboxing

---

1. Introduction & Motivation

An increasingly digitized or smart world has created dangerous times. Incidents related to cyberweapons like BlackEnergy [1] started as a simple distributed denial of service (DDoS) platform to a quite sophisticated plug-in based malware and Stuxnet [2]. This included the destruction of equipment in the operational technology (OT) environment which is a reminder about the vulnerabilities of this digital world. The impacts of such cyber-physical attacks on critical infrastructures resonate far beyond the confines of Iran and Ukraine. These were very well coordinated attacks, well-structured and pre-planned. In today's multi-polar world, it is foreseeable to consider that such types of attacks would continue to prosper with increased complexity.

Hence, in such a volatile world, it is of paramount importance to be well-prepared in terms of threat analysis and collection of digital evidence in order to attribute attacks to state and non-state actors. Forty-six cyber-attacks incidents were reported in the energy sector in 2015 [3], mostly related to the IT system of power utilities and their dealers. The U.S. Department of Energy indicates that the actual number of cyber-attacks is higher than reported [4].

The conventional electrical power grids are being transformed into smarter grids. Wide area monitoring protection and control (WAMPAC) systems, is one of the many advanced capabilities that are being equipped with the area of electric power systems. WAMPAC helps to improve planning, operation, and maintenance of electric grids [5]. The major components of this system are the phasor measurement units (PMUs) and the phasor data concentrators (PDCs). In a WAMPAC system, time-synchronized phasor data from multiple PMUs are integrated into a PDC, to produce a time-aligned output PDC data stream. A PDC is usually compliant with the IEEE C37.118 standard. PDC, being one of the most important parts of WAMPAC, is vulnerable to cyber-attacks.

Motivation

It is extremely important for a forensic investigation to get credible evidence for any cyber-attack attribution as required by the Daubert standard for it to be admissible in a court of law. Evidence should follow the Daubert standard and the integrity of the evidence must be flawless. If the integrity of evidence can be ensured and the Daubert standard is followed, only then can attributing evidence incriminating the culprits be presented. All of this only possible if there is a controller environment in which the parameters have been set and the type of the variations occurring with respective changes is known. In other words, the exact activity is mapped with exact attributing evidence.

A closed loop, non-interfering environment, such as sandboxing, helps to provide such an exact mapping. Previously, sandboxing has not been used in power systems, neither in digital forensics context, nor in security context. Thus, it perfectly makes sense to use such an important technique for evidence collection and analysis. All this clearly helps in ensuring the integrity of the acquired evidence, in turn, making a formidable case based on sound forensic artifacts.

2. Related Work

The use of digital forensic investigations in the domain of electric power systems is quite limited. The current research focus for digital forensics usually tends towards supervisory control and data acquisition (SCADA). For example, Ahmed et al. [6] discussed some measures for forensic readiness in the SCADA environment. Similarly, Wu et al. [7] discussed a SCADA digital forensic process consisting of seven steps and Eden et al. [8] discussed a SCADA forensic incident response model. A few other perspectives include works like Ahmed et al. [9] discussing programmable logic controller (PLC) forensics. Many works [10–14] explain the different types of cyber-attacks, testbeds and potential vulnerability in a digital substation in smart grids, however, they seldom discuss anything related to digital forensic investigations. For example, Stellios et al. [15] have discussed advanced persistent threats and zero-day exploits in the Industrial Internet of Things specifically detailing attacks on smart grid SCADA networks and field devices. These include the 2007 Aurora attack scenario that targeted electric power generators demonstrated by Idaho US National Labs [16], the 2015 attack on the Ukraine's smart grid distribution network [17], and the 2016 attack on the Ukraine's Kiev transmission station [18].

Among the first forensic investigations on intelligent electronic devices (IEDs) and phasor measurement units (PMUs) is the work done by Iqbal et al. [19]. They studied digital forensic readiness in industrial control systems (ICS), especially revolving around substation automation and devices called IEDs and PMUs as their case studies in smart grids. Through these case studies, they performed different attacks and tried collecting evidence. It was concluded that current ICS devices, i.e. IEDs and PMUs, in substations were not forensic ready. Similarly, Iqbal et al. [20] in their work investigated different logs of industrial control systems (ICS) on a variety of devices using a variety of operational setups. They concluded that in industrial control systems logs are relatively less mature, hence leading to the inability of logs to ascertain incrimination and attribution of an attack. They suggested the logs be modified in contents to contain more information in aid of forensic investigations when deemed necessary.

Forensics investigations require a test setup to be properly configured in order to acquire useful forensic artifacts in case of a cyber-attack. In a power system domain, many testbeds are proposed in the literature for simulation attacks [10–14], however, most of them are not suitable for collecting useful artifacts to be used later in forensic analysis. Among the first testbeds proposed in [21] was suitable for conducting a forensic investigation and potentially collecting artifacts.

One of the challenges associated with the digital forensic investigation is identification, examination, and extraction of digital artifacts with the probative value [22]. The complexity of the digital environment makes this a tedious task if not done properly. Moreover, the process of finding artifacts should be automated. The challenge is usually overcome by using various suitable tools and techniques for this purpose. Sandboxing is one of the suitable techniques used in digital forensics for this purpose.

A sandbox environment is an isolated environment although originally used by programmers to test new code, has now become one of the useful techniques for security and forensic investigations [22]. Although sandboxing has been in use for many years in other domains, but to authors' best knowledge, it has rarely been used in power system testbeds. In [23], the authors have proposed the concept of sandboxing for this first time in this domain, however, the focus of the paper was not exploring the details. The paper identifies attack-based digital forensic evidences (DFEs) for WAMPAC systems). A correlation is performed based on acquired DFEs resulting from system application logs. In [24], intelligent analysis of digital evidence is performed in large scale logs in power system attributed to the attacks.

Paper Contributions

In [21], a forensic testbed was proposed for simulating cyber-attacks on WAMPAC applications. Although the first of its nature for forensic investigations in the power systems, the challenges encountered in the investigation process led to improvisation of our implementation using the sandboxing approach. Additional experiments were performed on top of the existing work which had previously performed by the authors.

In this paper, the concept of sandboxing on the hardware-in-the loop forensic testbed originally proposed in [21] has been introduced. An authentication attack between the communication of PMU and PDC was considered. The PDC system is particularly sandboxed, which provides a controlled environment, thus, facilitates in extracting the artifacts with forensic value. Hence, such artifacts help in the investigation process to trace back to the root cause of the cyber-attacks.

The remainder of the paper is organized as follows:

In Section 3, a substation architecture is presented identifying the vulnerabilities and potential artifacts, which gives a broader overview of forensic investigation in the field of the power system and smart grids. Section 4 provides the theoretical details of the sandboxing approach. Section 5 describes the details of the power system model used to emulate the behavior of the power system used for the testbeds. Section 6 contains the mapping of different attacks in WAMPAC applications. In Section 7, the two approaches of real-time hardware-in-the-loop forensic testbed are presented Finally, Section 8 presents the key conclusions of the paper.

3. Background on Substation Architecture

A substation consists of various entities including hardware and software components. These components may be subjected to various types of attacks. Figure 1 shows a broad overview of the architecture of a substation. The figure identifies most of the probable vulnerabilities of the cyber-attacks. In addition, key potential places are detected where useful forensics artifacts could be acquired. The study focuses on some of the major substation devices, in the context of IEC 61850 substation automation architecture, i.e. PMUs, IEDs and remote terminal units (RTUs), etc.

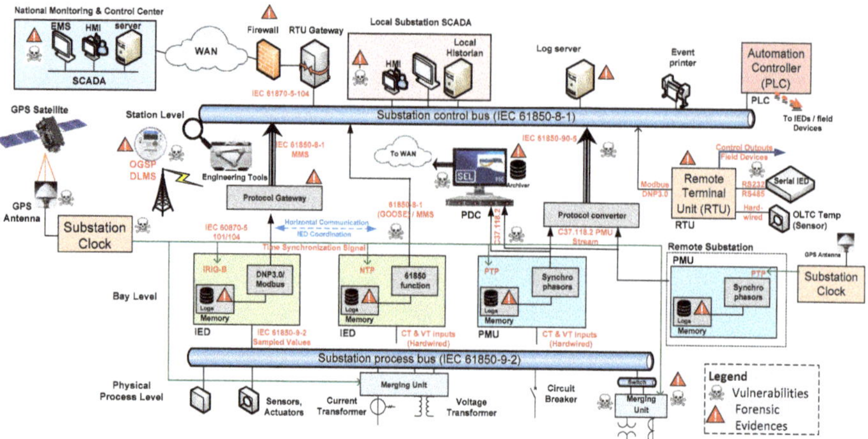

Figure 1. Substation architecture identifying vulnerabilities and forensic evidence.

As shown in Figure 1, data from physical processes are fed to the substation devices (i.e., IEDs and PMUs) at the bay level through the process bus. Based on this data, the IEDs perform various protection/control functions and stream out the status of substation digitally by using different power systems communication protocols, such as IEC 61850-8-1 (GOOSE/MMS) and IEC 60870-5 (101/104).

Protocol converters (protocol gateways) are deployed to map information between various communication protocols. This digital data set enables a station bus to facilitate various monitoring, logging and coordination functions within a substation.

An IED can either directly communicate with a station bus using 61850-8-1 GOOSE/MMS or requires a protocol gateway to publish data from the IED to the substation control bus (in case IED uses DNP3/Modbus or any other protocol which is not 61850 compliant). The data from IEDs and RTUs are sent to the national control center via the RTU gateway using IEC 61870-5-104, where SCADA/EMS system performs various monitoring and control functions.

The synchrophasors data complaint with C37.118.2 protocol streams out by the PMUs. [25]. The phasor data concentrator (PDC) receives data from various PMUs and streams out the time-aligned data via the wide area network (WAN) to a national control center, where the data is used by various monitoring, control, and protection applications. Moreover, a protocol converter parses PMU data (C37.118.2) protocol, maps it to the data model of IEC 61850 and sends the data through either the routed-sampled value or the routed-GOOSE services as per IEC 61850-90-5 protocol [26].

The major components of the substation architecture (shown in Figure 1) are explained briefly as follows:

(1) IED: An IED is a device having various monitoring, protection and control functionalities. It can be used for upper-level communication individualistically. The input to this device could be the data from the sensor and actuators from the field equipment. IED can issue various control orders, such as tripping circuit breakers to protect the network section and expensive equipment.

(2) PMU: A PMU is a device which estimates the synchrophasors, frequency, and rate of change of frequency (ROCOF) of the input voltage and /or current waveform, based on a common universal time (UTC) time reference. A PMU is usually installed into a power substation and connected to the electric grid via instrument transformers. PMUs have great benefits for situational awareness in WAMPAC and forensic event analysis [27]. For example, a PMU can report the data at a rate as high as up to 120 samples/ seconds. In addition, they have the capability to measure phase angle variations at multiple places which allow grid operators to detect and characterize the grid much faster than the traditional devices. A PMU measurement is time-stamped, and thus coordinated against universal time (UTC) using a time source, such as the GPS. The time synchronization feature of PMU measurements helps to improve the overall coordination of the grid operation.

(3) RTU: It is an integral part of typical SCADA systems used as a communication pivot. The data from sensors and actuators in substations and remote locations are collected by RTUs. It sends the collected data to a master station (national monitoring and control center) via a communication system. RTUs provide historical and sequence of event data for fault investigations and provide historical trend data for power network planning and maintenance.

(4) PDC: It receives time-synchronized phasor data from multiple PMUs to produce a real-time, time-aligned output data stream. A PDC can exchange phasor data with PDCs at other locations. A PDC is usually compliant with the IEEE C37.118 standard, which provides the required capability for the most advanced wide area monitoring applications. Moreover, a PDC also provides a gateway for the interface of PMU applications to the SCADA/EMS system to improve the supervision of the power system.

(5) Merging unit (MU): It is an IED that is used to exchange information between field devices (at process level) and secondary devices at the bay level. The MU collects the inputs from current and voltage transformers synchronously and provides output in the form of digital signals compliant with IEC 61850 protocol.

(6) Local Substation SCADA:

 a. Human machine interface (HMI): HMI is used by system operators and control center engineers to interpret and visualize SCADA system data through a graphical user interface. HMI can also be used for transferring algorithms, configuring set points and adjusting parameters of controllers.
 b. Historian and log servers: It is basically a database management system that acquires and saves data sent to the control center. It is also used to produce audit logs for all events and activities across a SCADA network. Therefore, it is a vital source of evidence for any incident forensic investigations.

(7) Energy management system (EMS) and SCADA: This is an energy management system that optimizes, supervises and controls the transmission grid and generation assets. It hosts all monitoring, control, protection, and planning applications, etc. The SCADA system can have built-in capabilities of EMS or it can be a separate system.

(8) Engineering tools: A set of software tools are the essential part of a modern substation automation architecture. The tools can be used to configure the bay level and process level devices. Moreover, the tools help in the overall system configuration. In addition, some appropriate tools could be used for a real-time cybersecurity monitoring of the substation.

(9) Time synchronization: This is required to ensure that all devices in a substation have accurate clocks for system control and data acquisition, etc. Time synchronization enables electric power utilities to have better coordination of the operation and helps to maintain power supply integrity.

(10) Protocol converters: Protocol converters parses data from one protocol, map it to another desired protocol and finally transmit the data accordingly. Having various protocols in substation architecture, protocol converters become important for the interoperability between different devices.

(11) Substation process bus (IEC 61850-9-2): It is an interface between primary field devices and secondary devices at the bay level. The process bus is compliant with IEC 61850-9-2 standard. Conventional instrument transformers provide analog values which can be exchanged for fiber-optic sensors. The modern non-conventional instrument transformers can communicate with the process bus via digital signals to be fed to metering, protection and control equipment.

(12) Substation station control bus (IEC 61850-8-1): It is an interface between secondary bay level devices and the control center/SCADA. The station bus is usually compliant with IEC 61850-8-1 standard.

4. Sandboxing

A sandbox environment is an isolated environment which is used to safely run suspicious codes without impacting the host. Although originally used by programmers to test the new code, it has now become one of the most useful techniques for security and forensic investigations especially for malware analysis [22]. Sandbox systems allow monitoring suspicious executable files while eliminating the risk of compromising live systems. Another important aspect is that sandboxes eliminate a lot of human effort derived from complex and lengthy tasks, such as disassembling [28]. There are several commercial and open source sandboxes available like Cuckoo Sandbox, Microsoft App-V, VMware ThinApp, Zerowine, etc. all with their pros and cons [29]. The concept of sandboxing has been well known for quite some time in other domains, but it has never been proposed or used with a hardware-in-the-loop (HIL) power system setup for digital forensics to the best knowledge of the authors. This study is using a modified real-time hardware-in-the-loop (HIL) system where some portions of the systems have been sandboxed [21].

There are several ways sandboxes can be implemented, for example, full system emulation, operating systems emulation, etc. In the first approach, a deep inspection is possible because all aspects of the hardware and software are analyzed, resulting in a fine-grained analysis. This is in contrast to the latter approach in which only software and user behavior are monitored. The second approach using Cuckoo Sandbox was chosen, essentially auditing user access, file storage, registry, connections, processes, etc.

A differential forensic analysis (DFA) was performed comparing two or more different digital forensic images and resulted in reporting the changes and modifications between them. By focusing on the changes and modifications, it allows the examiner to reduce the processing time and amount of information under examination. Also, focusing on changes helped segregate activities which potentially were performed by the malicious user [30].

Figure 2 depicts a DFA with an initial system state (SS_0) at an arbitrary t_0, taken as a base state of the system. After initiating an event (E_1), the base state SS_0 changes its state SS_1. All the changes occurring in the time interval (t_1-t_0) are defined as (Δ_1). In order to capture all such changes in the system states (Δ_n), the experiments were repeated several times in order to get a complete set of artifacts [23]. For the duration of our experiments, approximately 2 million logs over several iterations were collected. Since the volume of data, time and power required for processing were not in the scope of work, hence here no comparisons were presented.

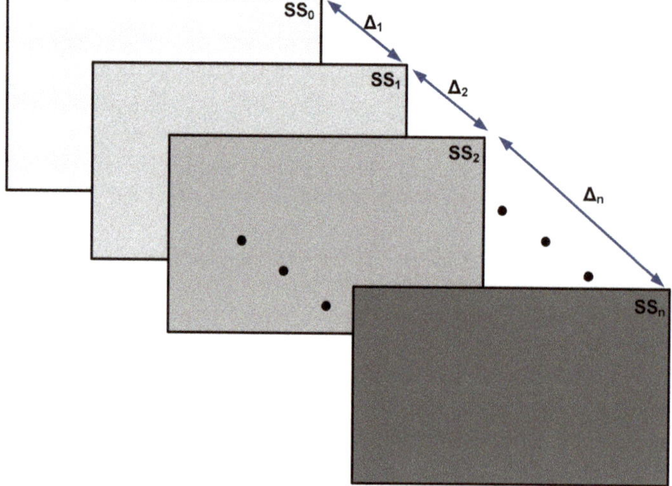

Figure 2. The concept of sandboxing and differential forensics analysis [23].

5. Power System Modeling

To emulate the behavior of the power systems, an IEEE 3-machine 9-bus [31] power system model is used as shown in Figure 3. The power system contains 3 generators available at three end buses (bus 1–3) of the network. The transformers are used to step up/down the voltage of the power network. Moreover, 3 loads (which consume power) are connected at bus 5, 6 and 8. The transmission lines are used to transfer power from generation points to the load centers. Each line, loads and generators are equipped with circuit breakers (CB), which gives the flexibility to only disconnect faults part of the power network. This increases the reliability of the network and gives improved availability of electric power.

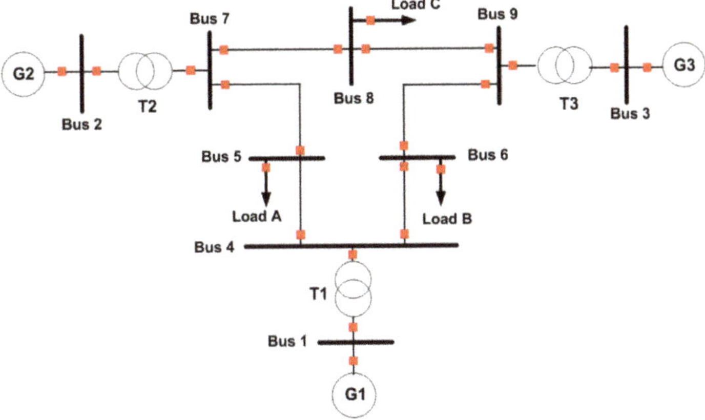

Figure 3. IEEE 3-machine 9 bus power system model.

The model was first designed for offline simulations in MATLAB/SIMULINK. The model was further prepared for real-time simulation using RT-LAB from Opal-RT real-time simulator (RTS). Figure 4a shows different steps of the offline model in Simulink prepared for its execution in the RTS. The Simulink model is first divided into sub-systems. The model is loaded into the RTS using the RT-LAB software interface. Finally, the model is executed in the RTS.

A detailed execution architecture for the real-time simulation of the model is shown in Figure 4b. The Simulink model can be broadly classified into computations part and the graphical user interface (GUI) part. The computation part is mostly executed in a real-time target, whereas the GUI part is executed in the local host computer.

In the real-time simulation, the build process contains the following steps. First, the Simulink model is divided into various subsystems for efficient computation purposes. It is then followed by C code generation, transfer of the generated code and finally building the generated code. After the build process, it is then followed by load process, in which the model is loaded into the real-time target as shown in Figure 4b. Once the model is loaded, it is finally executed. It should be noted that all the computation part is executed in real-time on the target, whereas all the graphics, including plots and display, is executed into the host computer.

In this paper, IEEE 9 bus model shown in Figure 3 is used to design, test and validate the attack simulation experiments. The model is an essential component of the forensic testbeds (shown in Section 7) to be used for simulation and analysis of cyber-attacks on substation devices.

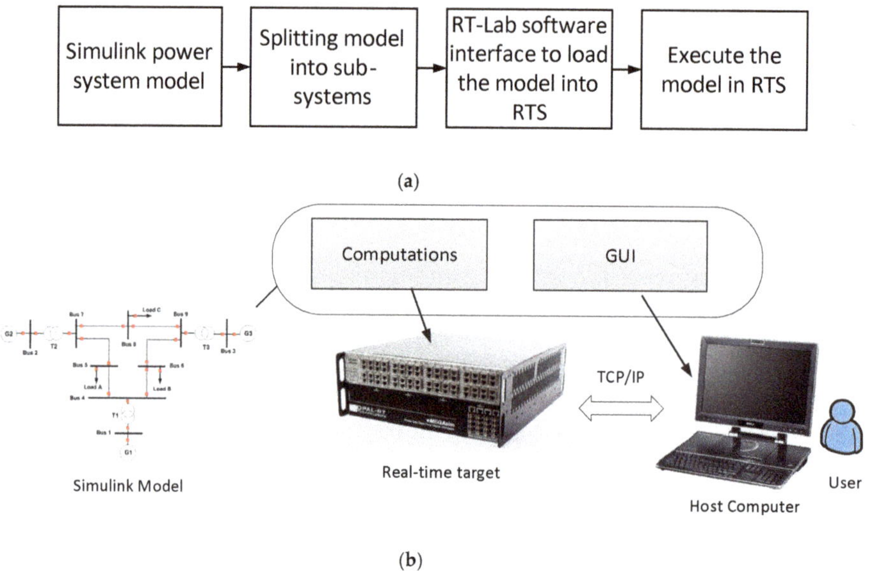

Figure 4. (**a**) Model preparation steps for real-time simulation; (**b**) Detailed execution architecture for real-time simulation.

6. Mapping of Different Attacks in WAMPAC Networks

In Figure 5, a comprehensive mapping was performed for different attacks in WAMPAC network. The focus was to map the attacks corresponding to potential forensic evidence that could be generated in a WAMPAC network. The mapping could help to get a quick understanding of the types of cyber-attacks, usually targeted resources, applications under attack and their potential impact on the power system.

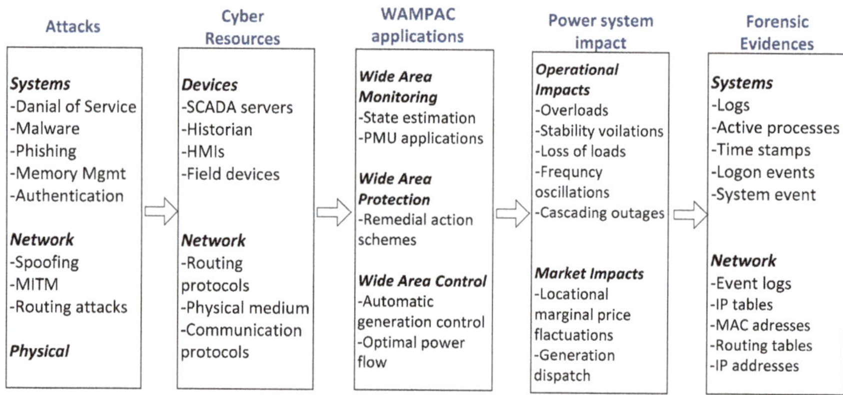

Figure 5. Mapping of different attacks in WAMPAC network with respect to potential forensic evidence [32].

The mapping is first of its kind in the field of WAMPAC applications with a specific focus on digital forensic evidence. The forensic evidence is broadly classified into system evidence and network evidence. This mapping provides a useful starting point for the digital forensics researchers to explore the applications of digital forensics into the field of power systems.

7. Real-time Hardware-in-the-Loop Forensic Experimental Testbed-Approaches

7.1. Without Sandboxing

A real-time HIL testbed was proposed in [21] as shown in Figure 6. The testbed is suitable for performing digital forensic experiments in the domain of the power system. A power system model was simulated as shown in Section 5. PMUs are fed by simulated voltage and current from the RTS. The synchrophasors are estimated by the PMUs in c37.118 data format. The data from multiple PMUs are collected and time-aligned by PDC, transmitting the data stream via WAN to the monitoring and control center. The monitoring, control and protection applications use PMU data and perform various functionalities accordingly. In this case, a time synchronization signal to the PMU is subjected to a cyber-attack.

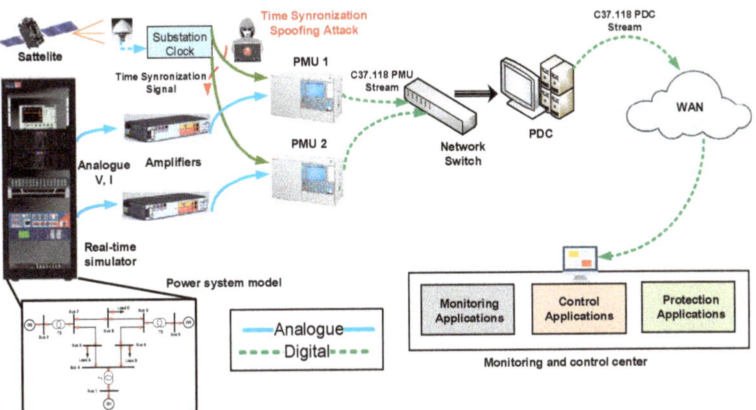

Figure 6. Forensic experimental testbed (Without Sandboxing).

Basically, the authors tried to retrace the steps of the time synchronization attacks, to find out potential forensic artifacts using the proposed testbed shown in Figure 6. However, due to the lack of a suitable mechanism, the evidence collection was manual in nature and was a tedious time-consuming task.

In Figure 7, a detailed sequence of time synchronization attack is shown. As the figure shows that a power system model is created, compiled, loaded and consequently simulated for both normal operation and operation under attack. In normal operation, healthy time synchronization signal is fed to PMUs, and PDC stream is being published to the target application to perform the relevant functions. However, in case of an attack, the attacker injects a spoofed time signal to a PMU and PDC publishes an infected stream to the target application which could negatively impact the performance of the target application.

The time synchronization input to a PMU is considered to be one of the most susceptible signals for a cyber-attack. An attack on time synchronization input signal infects PMU data, which could badly impact the end-user applications. For instance, a PMU based monitoring application shown in Figure 8, i.e., steady state model synthesis (SSMS) [33], produces reduced equivalent network models using PMU data. The attack on time-stamp could corrupt the calculations of the parameters of the equivalent network model. The system operator utilizing the output of the SSMS application could potentially take wrong decisions based on the corrupted results. This could result in unwanted interruptions in the power network and eventually could result in a blackout of the power network.

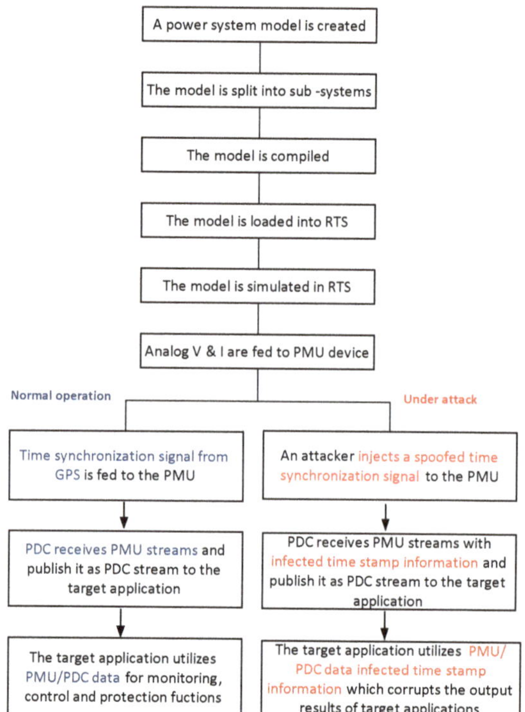

Figure 7. The sequence of operation of time synchronization attack.

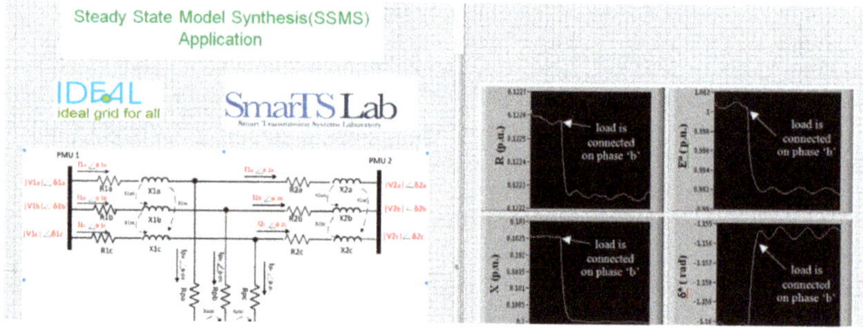

Figure 8. SSMS power system monitoring application.

7.2. With Sandboxing

7.2.1. Design of Sandboxed Forensic Experimental Testbed (Authentication Attack)

In this paper, a forensic experimental testbed is proposed using sandboxing approach as shown in Figure 9. The base design of this testbed is similar to the testbed as described in Section 7.1. The most predominant difference is that the PDC system is sandboxed.

In this testbed, the PDC is subjected to an authentication attack. It is assumed that the attacker has gained unauthorized access to the PDC network and to the PDC's credentials. After gaining access to the credentials of the PDC system, the attacker performs various unwanted malicious activities on the PDC system.

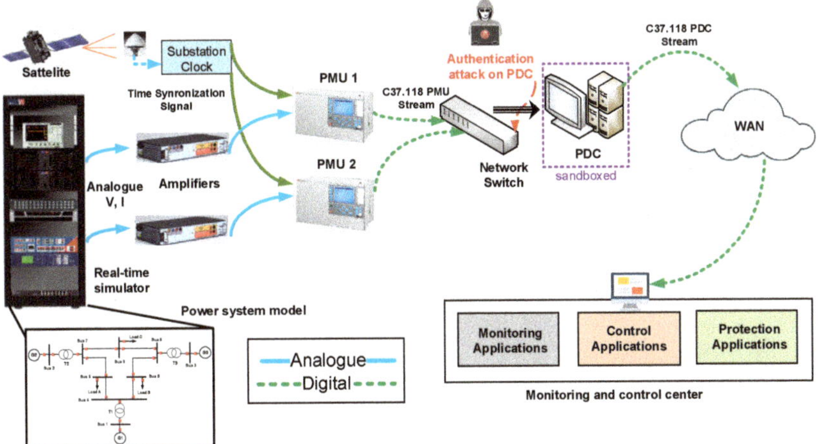

Figure 9. Forensic experimental testbed (With Sandboxing), authentication attack [23].

7.2.2. Authentication Attack Steps

Figure 10 shows the timeline of the list of activities performed by the attacker on the PDC system. The time of occurrence of various activities are plotted on the x-axis and the list of activities are taken on the y-axis. As shown in the figure, the monitoring of PDC started at 16:45. Soon after, 2 minutes, PDC services and PDC graphical assistance software started. The attacker performed a couple of failed attempts to login into the PDC system at 16:50 and 16:52 consecutively. The attacker was able to successfully login into the PDC system at 16:54. At 16:56, the PDC stream was disabled by the attacker, causing a denial of service to the PDC stream. The attacker enabled the PDC stream back in operation after 2 minutes of the attack. The attacker logged out of the PDC assistance software and closed the PDC services at 17:00 and 17:01 consecutively.

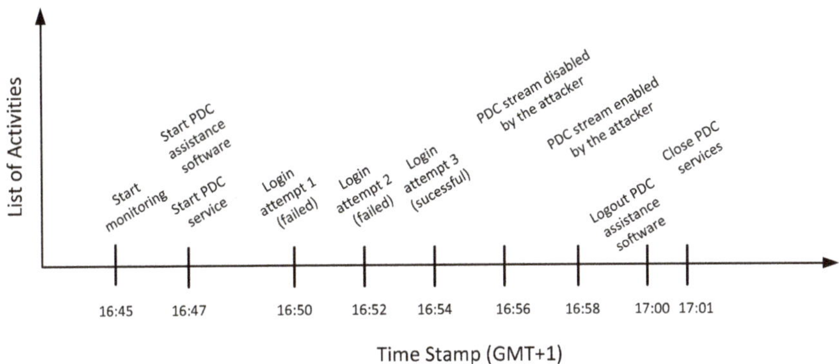

Figure 10. Timeline analysis of the activities performed.

7.2.3. Sandboxing on PDC System

The architecture of the PDC system is shown in Figure 11. As shown in the figure, C37.118 2005 clients communicate with the PDC C37.118 2005 server. The PDC first collects the data from different PMUs, processes the data by time aligning and it finally broadcasts it in the form of the data stream. The database in the PDC system logs all the events. The configurations and settings of PMU/ PDC streams by the users are performed using graphical user interface assistance software on a local computer. Whereas, once the configurations are saved, the updated data is transferred to the PDC server.

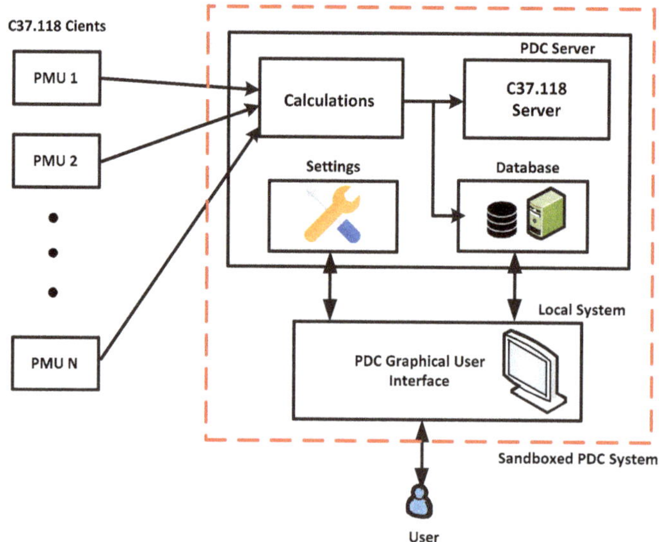

Figure 11. The architecture of a PDC system with sandboxing.

As shown in Figure 11, the PDC system, i.e., server and GUI, are sandboxed. All the attack steps performed by the attackers on the PDC system generated some potential digital artifacts on the PDC system. For example, the evidence generated as a result of most of the malicious activities shown in the timeline in Figure 10 can be associated with their respective steps. The extraction of these useful artifacts could only be made possible due to a controlled sandboxed, which keeps track of each minor change on the PDC system. These recorded changes provide vital input for a successful forensic investigation.

8. Conclusions

The paper proposes the design of a real-time hardware-in-the-loop forensic experimental testbed to conduct investigations on power systems as a whole. This is quite a unique work in its own since nothing comparable existed before [21]. The proposed test setup is used for forensic analysis of substation devices particularly focusing on WAMPAC applications. As a result of cyber attacks, constituent devices leave the traces behind known as forensic artifacts as per Locard's principle. These forensic artifacts were identified, preserved, collected and studied to know how exactly such attacks impacted the system and what traces they left behind.

Secondly, based on the challenges and difficulties faced in conducting such investigations, certain improvements were made specifically adding and implementing the sandboxing technique. Sandboxing has been used frequently by programmers in the past, and now-a-days is considered very important while performing security analysis. It has proven to be very useful as an aid in forensic analysis but is rarely used in the context of industrial control systems.

This study successfully implemented the sandboxing technique in RT-HIL for conducting forensic analysis. Problems such as volume of data involved and tracking different kinds of changes in the file systems which not only took more time, but also was tedious in nature and time-consuming task. Sandboxing helped in reducing these difficulties considerably. While previously, the authors were only able to perform and analyze time synchronization and GPS spoofing attack in this work, we expanded our horizons by adding authentication attack. These attacks resulted in successful identification and collection of more forensic artifacts.

In future work, the authors potentially see the opportunity to perform different attacks. This can include different types of attacks on the same device to find out which attacks leave more artifacts and which attacks leave less artifacts. This may also be useful for mapping all the relevant artifacts to different attack types. Another possible direction is to increase the scope of sandboxing which may be helpful in conducting a wider collection and analysis of artifacts.

Author Contributions: The corresponding author, A.I. and co-author, F.M. have both performed the experiments and wrote the paper whereas co-author, M.E. has helped through reviewing and feedback.

Funding: This work has received funding from the Swedish Civil Contingencies Agency (MSB) through the research center Resilient Information and Control Systems (RICS).

Acknowledgments: Authors would like to thank Lars Nordström Head of EPE Department, Deputy Head School of EECS, KTH and Gunnar Björkman of ABB for their support.

Conflicts of Interest: The authors declare no conflicts of interest.

Abbreviations

WAMPAC	Wide Area Monitoring Protection and Control
IED	Intelligent Electronic Device
PMU	Phasor Measurement Device
RTU	Remote Terminal Units
PDC	Phasor Data Concentrator
RTS	Real-time Simulator

References

1. Khan, R.; Maynard, P.; McLaughlin, K.; Laverty, D.M.; Sezer, S. Threat Analysis of Black Energy Malware for Synchrophasor based Real-time Control and Monitoring in Smart Grid. In Proceedings of the ICS-CSR, Belfast, UK, 23–25 August 2016.
2. Langner, R. Stuxnet: Dissecting a cyberwarfare weapon. *IEEE Secur. Priv.* **2011**, *9*, 49–51. [CrossRef]
3. NCCIC; ICS-CERT. *NCCIC/ICS-CERT 2015 Year in Review*; NCCIC; ICS-CERT: US Department of Homeland Security: Washington, DC, USA, 2016.
4. The US Department of Energy. *Cyber Threat and Vulnerability Analysis of the US Electric Sector*; US Department of Energy: Washington, DC, USA, 2016.
5. Terzija, V.; Valverde, G.; Cai, D.; Regulski, P.; Madani, V.; Fitch, J.; Skok, S. Wide-Area Monitoring, Protection, and Control of Future Electric Power Networks. *Proc. IEEE* **2011**, *99*, 80–93. [CrossRef]
6. Ahmed, I.; Obermeier, S.; Naedele, M.G.G.R. SCADA Systems: Challenges for Forensic Investigators. *Computer* **2012**, *45*, 44–51. [CrossRef]
7. Wu, T.; Pagna Disso, J.F.; Jones, K.; Campos, A. Towards a SCADA forensics architecture. In Proceedings of the 1st International Symposium for ICS SCADA Cyber Security Research, Leicester, UK, 16–17 September 2013.
8. Eden, P.; Blyth, A.; Burnap, P.; Cherdantseva, Y.; Jones, K.; Soulsby, H.; Stoddart, K. A cyber forensic taxonomy for SCADA systems in critical infrastructure. In Proceedings of the International Conference on Critical Information Infrastructures Security, Berlin, Germany, 5–7 October 2015.
9. Ahmed, I.; Obermeier, S.; Sudhakaran, S.; Rousse, V. Programmable Logic Controller Forensics. *IEEE Secur. Priv.* **2017**, *15*, 1518–1524. [CrossRef]
10. Sun, C.C.; Hahn, A.; Liu, C.C. Cyber security of a power grid: State-of-the-art. *Int. J. Electr. Power Energy Syst.* **2018**, *99*, 45–56. [CrossRef]
11. Xu, Y.; Yang, Y.; Li, T.; Ju, J.; Wang, Q. Review on cyber vulnerabilities of communication protocols in industrial control systems. In Proceedings of the 2017 IEEE Conference on Energy Internet and Energy System Integration (EI2), Beijing, China, 26–28 November 2017.
12. Leger, A.S.; Spruce, J.; Banwell, T.; Collins, M. Smart grid testbed for Wide-Area Monitoring and Control systems. In Proceedings of the 2016 IEEE/PES Transmission and Distribution Conference and Exposition (T&D), Dallas, TX, USA, 21–23 September 2016.

13. Zhong, X.; Jayawardene, I.; Venayagamoorthy, G.K.; Brooks, R. Denial of Service Attack on Tie-Line Bias Control in a Power System with PV Plant. *IEEE Trans. Emerg Top. Comput. Intell.* **2017**, *1*, 375–390. [CrossRef]
14. Calvo, I.; Etxeberria-Agiriano, I.; Iñigo, M.A.; González-Nalda, P. Key vulnerabilities of industrial automation and control systems and actions to prevent cyber-attacks. *Int. J. Online Eng. (IJOE)* **2016**, *12*, 9–16. [CrossRef]
15. Stellios, I.; Kotzanikolaou, P.; Psarakis, M. Advanced Persistent Threats and Zero-Day Exploits in Industrial Internet of Things. In *Security and Privacy Trends in the Industrial Internet of Things. Advanced Sciences and Technologies for Security Applications*; Alcaraz, C., Ed.; Springer: Cham, Switzerland, 2019.
16. Srivastava, A.; Morris, T.; Ernster, T.; Vellaithurai, C.; Pan, S.; Adhikari, U. Modeling cyber-physical vulnerability of the smart grid with incomplete information. *IEEE Trans. Smart Grid* **2013**, *4*, 235–244. [CrossRef]
17. Case, D.U. *Analysis of the Cyber-Attack on the Ukrainian Power Grid*; Electricity Information Sharing and Analysis Center (E-ISAC): Washington, DC, USA, 2016.
18. Goodin, D. Hackers Trigger Yet Another Power Outage in Ukraine. Available online: https://arstechnica.com/security/2017/01/the-new-normal-yet-another-hacker-caused-power-outage-hits-ukraine/ (accessed on 18 May 2019).
19. Iqbal, A.; Ekstedt, M.; Alobaidli, H. Digital Forensic Readiness in Critical Infrastructures: A case of substation automation in the power sector. In Proceedings of the International Conference on Digital Forensics and Cyber Crime, Prague, Czech, 9–11 October 2017.
20. Iqbal, A.; Ekstedt, M.; Alobaidli, H. Exploratory studies into forensic logs for criminal investigation using case studies in industrial control systems in the power sector. In Proceedings of the 2017 IEEE International Conference on Big Data (Big Data), Boston, MA, USA, 11–14 December 2017.
21. Iqbal, A.; Mahmood, F.; Ekstedt, M. An Experimental Forensic Testbed: Attack-based Digital Forensic Analysis of WAMPAC Applications. In Proceedings of the Mediterranean Conference on Power Generation, Transmission, Distribution and Energy Conversion (MedPower 2018), Dubrovnik, Croatia, 12–15 November 2018.
22. Iqbal, A.; Al Obaidli, H.; Guimaraes, M.; Popov, O. Sandboxing: Aid in digital forensic research. In Proceedings of the 2015 Information Security Curriculum Development Conference (InfoSec '15), New York, NY, USA, 10 October 2015.
23. Iqbal, A.; Mahmood, F.; Shalaginov, A.; Ekstedt, M. Identification of Attack-based Digital Forensic Evidences for WAMPAC Systems. In Proceedings of the 2018 IEEE International Conference on Big Data (Big Data), Seattle, WA, USA; 2018; pp. 3079–3087.
24. Iqbal, A.; Shalaginov, A.; Mahmood, F. Intelligent analysis of digital evidences in large-scale logs in power systems attributed to the attacks. In Proceedings of the 2018 IEEE International Conference on Big Data (Big Data), Seattle, WA, USA, 10–13 December 2018.
25. IEEE. *IEEE Standard for Synchrophasor Data Transfer for Power Systems*; IEEE Std C37.118.2-2011 (Revision of IEEE Std C37.118-2005); IEEE: Piscataway, NJ, USA, 2011; pp. 1–53.
26. Firouzi, S.R.; Vanfretti, L.; Ruiz-Alvarez, A.; Mahmood, F.; Hooshyar, H.; Cairo, I. An IEC 61850-90-5 gateway for IEEE C37.118.2 synchrophasor data transfer. In Proceedings of the 2016 IEEE Power and Energy Society General Meeting (PESGM), Boston, MA, USA, 21–27 July 2016.
27. Mahmood, F. Synchrophasor Based Steady State Model Synthesis of Active Distribution Networks. KTH, 2018. Available online: http://kth.diva-portal.org/smash/get/diva2:1223943/FULLTEXT01.pdf (accessed on 18 May 2019).
28. Vasilescu, M.; Gheorghe, L.; Tapus, N. Practical malware analysis based on sandboxing. In Proceedings of the 2014 RoEduNet Conference 13th Edition: Networking in Education and Research Joint Event RENAM 8th Conference, Chisinau, Moldova, 11–13 September 2014.
29. Kale, G.; Bostanci, E.; Çelebi, F. Differences between Free Open Source and Commercial Sandboxes. In Proceedings of the International Conference on Cyber Security and Computer Science, Safranbolu, Turkey, 18–20 October 2018.
30. Garfinkel, S.; Nelson, A.J.; Young, J. A general strategy for differential forensic analysis. *Digit. Investig.* **2012**, *9*, S50–S59. [CrossRef]
31. Aggarwal, G.; Mittal, A.; Mathew, L. MATLAB/Simulink Model of Multi-machine (3-Machine, 9-Bus) WSCC System Incorporated with Hybrid Power Flow Controller. In Proceedings of the 2015 Fifth International Conference on Advanced Computing & Communication Technologies, Haryana, India, 21–22 February 2015.

32. Ashok, A.; Govindarasu, M.; Wang, J. Cyber-Physical Attack-Resilient Wide-Area Monitoring, Protection, and Control for the Power Grid. *Proc. IEEE* **2017**, *105*, 1389–1407. [CrossRef]
33. Mahmood, F.; Hooshyar, H.; Lavenius, J.; Bidadfar, A.; Lund, P.; Vanfretti, L. Real-Time Reduced Steady-State Model Synthesis of Active Distribution Networks Using PMU Measurements. *IEEE Trans. Power Deliv.* **2017**, *32*, 546–555. [CrossRef]

Article

Comparison of Measured and Calculated Data for NPP Krško CILR Test

Štefica Vlahović *, Siniša Šadek, Davor Grgić, Tomislav Fancev and Vesna Benčik

University of Zagreb Faculty of Electrical Engineering and Computing, Unska 3, Zagreb 10000, Croatia; sinisa.sadek@fer.hr (S.Š.); davor.grgic@fer.hr (D.G.); tomislav.fancev@fer.hr (T.F.); vesna.bencik@fer.hr (V.B.)
* Correspondence: stefica.vlahovic@fer.hr; Tel.: +385-1-6129-997

Received: 10 May 2019; Accepted: 5 June 2019; Published: 7 June 2019

Abstract: Containment is the last barrier for release of radioactive materials in the case of an accident in the nuclear power plant (NPP). Its overall integrity is tested during a containment integrated leak rate test (CILRT) at the design pressure, at regular intervals. Due to applied risk based licensing, the test intervals can be increased up to once in 10 years and beyond. Taking that into account it is important to prepare the test properly and to use obtained results to assess the real status of the containment. The test can be used to verify existing containment calculation models. There is a potential benefit of verified computer models usage for the explanation of some test results, too. NPP Krško has performed CILRT during the plant outage in 2016. The paper presents a comparison between measured data and results calculated using a multivolume GOTHIC (Generation Of Thermal Hydraulic Information For Containment) model. The test scenario was reproduced using limited available data up to the end of the pressurization phase. The depressurization phase is calculated by the code and measured leakage rate is implemented in the model. Taking into account the necessary adjustments in the model, overall prediction of the measured results (in terms of pressure, temperature and humidity) is very good. In the last phase of the test some non-physical behavior is noticed (without influence on overall test results), probably caused by the combination of air redistribution within the containment and influence of heat transfer to plant systems that were in the operation during the test. GOTHIC model was used to check sensitivity of the predicted pressure (leak rate) to different heat inputs and to investigate the influence that operation of only one reactor containment fan cooler (RCFC) train during pressurization can have on the mixing of air within the containment. In addition, the influence of currently used weighting factors (weighting of measured temperature, relative humidity and pressure data) on the used test methodology is investigated. The possible non-conservative direction of the influence (currently used weighting factors are giving lower leakage rate) was demonstrated and a new set of weighting factors is proposed too.

Keywords: NPP Krško; CILRT; GOTHIC code; containment; mass point method measurement; weighting factors

---

1. Introduction

Containment leakage should be kept at the minimum rate in order to restrict possible radioactive emissions in the environment below acceptable limits. The maximum allowable containment leakage rate is defined at the peak containment internal pressure (410 kPa) caused by the design basis accident (DBA). In nuclear power plant (NPP) Krško (NEK) the design rate is 0.2% of air weight per day [1]. This value is usually denoted as L_a and is equivalent to a leakage rate of 3.75×10^{-3} m^3/s [2].

Regular testing is required to ensure the leak rate limits are not exceeded. There are three types of tests in the containment leakage rate testing program: Type A, B and C. Type A test is actually the containment integrated leak rate test (CILRT) intended to measure the reactor containment overall integrated leakage rate under conditions representing design basis loss of coolant accident (LOCA) peak pressure (410 kPa). Type B tests are intended to pneumatically detect and measure local leakage across each pressure retaining or leakage limiting boundary for containment mechanical and electrical penetrations. Type C tests are intended to pneumatically measure containment isolation valve leakage rates. The purpose of all tests is to ensure that leakage through the containment or systems and components penetrating the containment does not exceed allowable leakage rates specified in the NEK Technical Specifications. Integrity of the containment structure is maintained during its service life.

NPP Krško CILRT test was performed in October 2016 during the regular outage. Based on temperature, pressure and relative humidity (RH) measurements, the leakage rate was calculated in accordance with standard ANSI/ANS-56.8-2002 [3]. In order to independently verify test data, the existing GOTHIC plant model [4] was adapted to comply with the test alignment and conditions, and was used to simulate containment behavior during the CILRT. Both the measured and calculated results are presented and evaluated in paper.

2. Containment CILRT Test

Pressurizing a large structure, such as the containment with a free volume of about 40,000 m^3, is not an easy task. Careful planning, setting up containment systems and equipment in required positions, venting and draining connecting lines, preparing the testing equipment according to standards [3], etc. are prerequisites for this kind of work.

The test is divided in five phases:

1. Pressurization phase,
2. Stabilization phase,
3. Testing phase,
4. Verification phase,
5. Depressurization phase.

Before the test it was necessary to prepare the containment in a way to determine the local leakage rate using local leakage rate tests of types B and C. The pressure, temperature, relative humidity, dew temperature and volume flow were measured during the test.

During the pressurization phase, which lasted 18 h, the test pressure 408 kPa (absolute pressure) was achieved. The average pressurization rate was 17 kPa/h. Five mobile compressors connected to the CILRT test containment penetrations were used in that process. During the stabilization phase, which lasted 4 h, containment conditions were controlled in parallel with preparations for the testing phase. The intention of that phase was to obtain stable containment initial conditions (pressure, temperature) in order to obtain correct mass leakage rate measurements in the subsequent testing phase. The reactor containment fan cooler (RCFC) Train A was available during pressurization and stabilization phases.

The minimal time interval for the stabilization phase is 4 h. After that time interval, all criteria according to ANSI/ANS-56.8-2002 standard were obtained.

The testing phase lasted 8 h. The aim of this phase was to evaluate containment inventory mass decrease. Based on temperature and pressure measurements, the air mass was calculated using the ideal gas law:

$$m = \frac{p \cdot V}{R \cdot T}. \tag{1}$$

The processing of measured data was carried out by the plant, according to ANSI/ANS-56.8-2002 standard, using three standard procedures (mass point method, total time method and point-to-point method) for calculating containment leakage rate. The procedures were implemented in the Excel spreadsheet. There were 24 temperature measurements sensors installed throughout the containment and two pressure sensors, in addition to six relative humidity measurements and two verification flow measurements. The average containment temperature was calculated using measured temperatures and the weighting factors depending on the free volume of the area where the sensor is installed.

The maximum allowable leakage by the ANS (American Nuclear Society) standard is 0.75 L_a, which is equal to 0.15% of air weight loss per day.

Two criterion types are used to evaluate the testing phase: The as-found and as-left leakage rate requirements.

The as-found leakage rate is the leakage rate prior to any repairs or adjustments to the containment barrier being tested. The criterion is defined as:

$$UCL + \Delta L + TLS < L_a, \tag{2}$$

where UCL is the statistical upper confidence limit of the measured data, ΔL is the leak penalties for penetrations not prepared for CILRT due to a number of reasons, such as to keep the plant in safe condition during the test, to provide cooling in the case of an accident, etc. and TLS is the total leakage savings for penetrations with initial high leakage that were repaired after the test. The sum of these three terms was about 20% of maximum leakage L_a.

The as-left criterion prescribes the total leakage made of the upper confidence limit and leak penalties to be less than 75% of the maximum allowable leaking rate L_a:

$$UCL + \Delta L < 0.75 \cdot L_a. \tag{3}$$

This limit does not include leakage savings since the impaired penetrations are repaired after the test. The as-left criterion was measured to be 25% of the prescribed leakage of 0.75 L_a.

In the testing phase final hour (the testing phase lasted 8 h), some non-physical behavior was noticed (increase of air mass in the containment in the situation when the containment is at higher pressure than any surrounding building or environment). It is expected that some combination of air redistribution within the containment and influence of the heat transfer to plant systems that were in the operation during the test can cause such behavior. That result should be further studied in more detail, but, in overall, the testing phase proved that the containment is airtight and no significant radioactive releases could be expected in a case of a DBA or severe accident causing increased containment pressure.

The following verification phase, which lasted 4.5 h, was conducted in order to check the accuracy of the testing phase results. The superimposed leakage rate method was used to verify the containment air mass decrease calculation. A valve connecting the containment to the outside atmosphere was opened and the leakage was set to 3.083×10^{-3} m^3/s. According to specifications [3] that superimposed leakage rate needs to fit between 0.75 L_a and 1.25 L_a. The flow rate of 3.083×10^{-3} m^3/s corresponds to 0.82 L_a. The same method as in the previous phase was used to calculate the mass inventory. The results confirmed the accuracy of the mass calculations. The calculated mass in the containment decreased with the same rate as was the leakage through the valve. The acceptability criterion for the total verification phase leakage rate L_c is:

$$L_0 + L_{tc} - 0.25L_a < L_c < L_0 + L_{tc} + 0.25L_a. \tag{4}$$

The flow rate L_0 corresponds to the superimposed verification flow rate and L_{tc} to calculate the leak rate in the testing phase. Thus, the lower limit leak rate calculated in such way is 0.73 L_a and the upper limit 1.23 L_a. The calculated leak rate in the verification phase varied slightly with the mean value 0.8 L_a. The acceptability criterion was satisfied.

The final depressurization phase lasted 15 h. A controlled air release, the release rate was carefully monitored in order not to damage sensitive containment equipment, was performed to depressurize the containment to standard atmospheric conditions.

3. GOTHIC Model of the NPP Krško

The computer code GOTHIC [5,6] was used for calculation of containment thermal hydraulic behavior. The code solves the conservation of mass, momentum and energy equations for multiphase (vapor phase, continuous liquid phase and droplet phase) multicomponent (water, air, hydrogen and noble gases) compressible fluid flow. In order to predict the interaction between phases for non-homogenous, non-equilibrium flow, the constitutive relations were used. One dimensional finite difference discretization was used to model heated or unheated structures. Hydraulic volumes can be lumped or subdivided in 1, 2 or 3 dimensions. Special models were used to simulate the operation of engineered safety equipment: Pumps, fans, valves, doors, vacuum breakers, spray nozzles, heat exchangers, heaters and coolers. Each component can be controlled by trip logic based on time, pressure, vapor temperature, liquid level or conductor temperature. The version of the code used in calculations was 7.2b. The newer version (version 8.3) was available, but for this type of the problem both versions are expected to give the same results.

NPP Krško containment outline is shown in Figure 1.

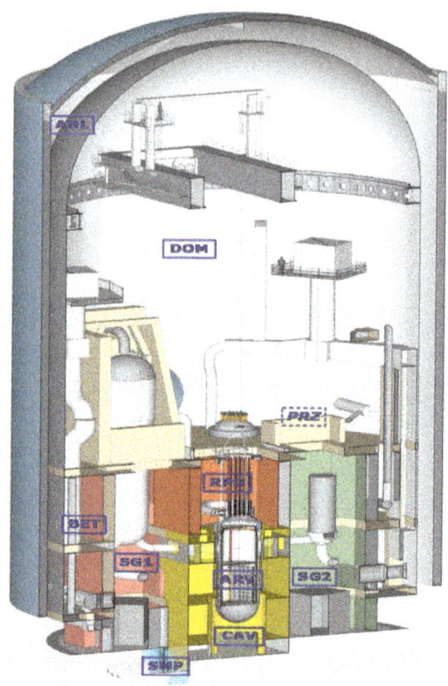

Figure 1. Nuclear power plant (NPP) Krško containment outline.

The GOTHIC multi compartment model of the containment was developed based on the plant's SAR (safety analysis report) Chapter 6 licensing model [7] and verified against the vendor's DBA LOCA analysis [4]. Figure 2 is taken directly from the GOTHIC code and shows the basic layout of used volumes and complexity of used internal flow paths and heat structures. In order to perform this calculation, the existing model is slightly modified. The initial conditions were adjusted to take into account real conditions in the plant in the beginning of the test. The flow path and the flow boundary condition were added to simulate the pressurization of the containment. However, the problem was that the plant personnel had not implemented a flow measurement for air added by the compressors. There was neither a measurement of the temperature nor RH of added air. The initial mass flow rate and thermodynamic parameters of the compressed air during pressurization were adjusted to obtain the measured rate of containment pressure and temperature increase. An iterative approach was used to reproduce initial phase of the test (during the pressurization) and to determine leakage flow path cross section (in the testing phase). A leak path cross section was selected to get a measured pressure behavior during the testing phase ($A = 10^{-6}$ m^2, $D_e = 1.128 \times 10^{-3}$ m, which is much less than the value usually used to model the containment design leak rate). Calculation of the verification phase was based on the cross section of the pipe and the measured discharge flow rate (real valve position was not known). The conditions during the last part of the test, the depressurization phase, were approximate. The plant personnel wanted to perform fast depressurization within allowable pressure decrease limits. Similarly, the calculation of that phase was just an attempt to follow a similar approach.

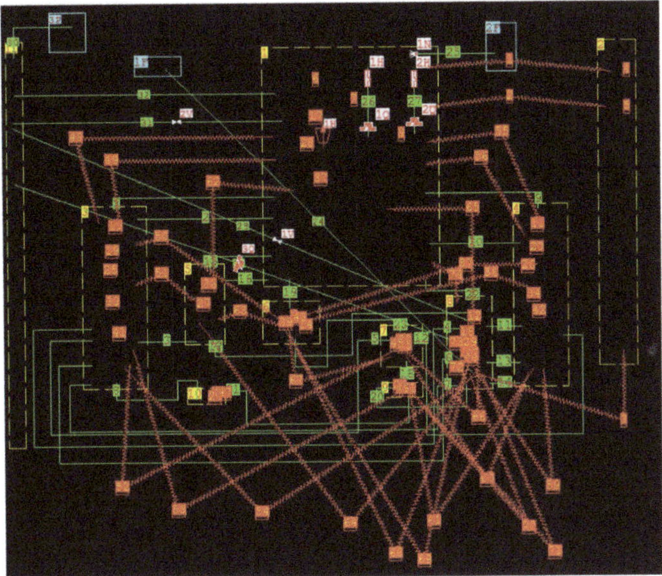

Figure 2. GOTHIC containment nodalization.

The containment was modeled with 10 hydraulic volumes representing the following compartments:

1. Containment dome (DOM),
2. Annulus (ANL),
3. Steam generator (SG) 1 compartment (SG1),
4. SG2 compartment (SG2),
5. Pressurizer (PRZ),
6. Refueling pool (RPO),
7. Space around reactor vessel (ARV),
8. Space around equipment compartments—containment lower plenum (BET),
9. Reactor cavity (CAV),
10. Containment sump (SMP).

There are 74 heat structures in the model. Steel liner, shield building wall, internal concrete walls and floors were explicitly modeled. Other internal heat structures (mostly carbon and stainless steel) were based on the SAR Chapter 6 containment design data. The heat structures were divided between control volumes depending on their approximate location. The internal heat structures were mainly heat capacities in the model so it was decided to preserve their mass and area exposed to the compartment. For bounding heat structures (both to the annulus and between compartments) the first priority was to determine the proper thickness and heat transfer area of the structures. The heat transfer coefficients were calculated internally by the code. Different combinations of Uchida condensation coefficients and natural heat transfer coefficients were used depending on the orientation of heat structures.

The flow paths in the model were based on real openings and communications between the compartments. More than one opening was used between the same volumes if they were located at different elevations to promote internal thermal mixing flow. That can be important for long-term containment transients. When required, e.g., in case of numerical difficulties, it is possible to join more flow paths in one equivalent flow path. The leakage (initially based on the design leakage) flow paths exist between the containment dome and annulus and between the annulus and the environment. In case of CILRT, the annulus was open to the environment and we were simply interested in the equivalent leak of the containment liner and that was the reason why we used a direct connection of the containment and environment (instead of the usual two step connection). Another reason for using a direct connection was that during containment depressurization a dedicated line was used to decrease the containment pressure by discharging containment air to the environment. In the model we used two parallel flow paths of different cross section areas (the discharge path had a valve and the leakage path was without it). The leak path is a connection of a very small cross section and it represents the whole containment leakage. It is possible, in a real situation, to have distributed the leak, but it is usual to address it with a concentrated leak path. GOTHIC has a separate model to address distributed leakage, however it was not used here because we wanted to keep connection with other codes (e.g., MAAP—Modular Accident Analysis Program) where the leak is expressed in terms of simple cross section area. All internal ventilation ducts can be taken into account (not used in this calculation), what will result in an increase of used control volumes beyond the number used for main compartments. In addition, the nodalization can be further subdivided axially using floor elevations and laterally in two halves taking into account existence of two RCFC trains and giving us opportunity to study the influence of a single train RCFC operation on the achieved initial conditions before the test. The additional subdivision of the containment was not used in this calculation, but it is planned for next analyses.

4. Results and Discussion

4.1. GOTHIC Calculation and Comparison With the Test Data

Containment pressure and temperatures are shown in Figures 3–5. Calculated values were in good agreement with the measured data. Pressurization, measurement and depressurization phases were clearly visible. Pressure scale during testing and verification phases was magnified to highlight the pressure decrease caused by leakage and the testing procedure (Figure 4). During the verification phase, an additional flow path between the containment and the environment was opened to calculate and verify the testing phase results. The calculated temperature decrease (label tv1_calc, other labels are used for representative locations measured temperature) in the depressurization phase was much faster than the measured data (Figure 5), due to the already mentioned approximate nature of the calculation in that phase (uncertainty of the boundary conditions—unknown depressurization rate).

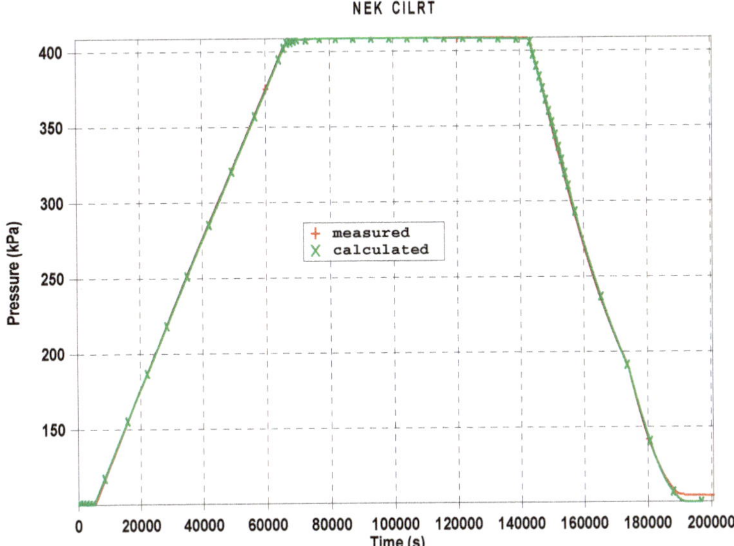

Figure 3. Measured and calculated containment pressures during the test.

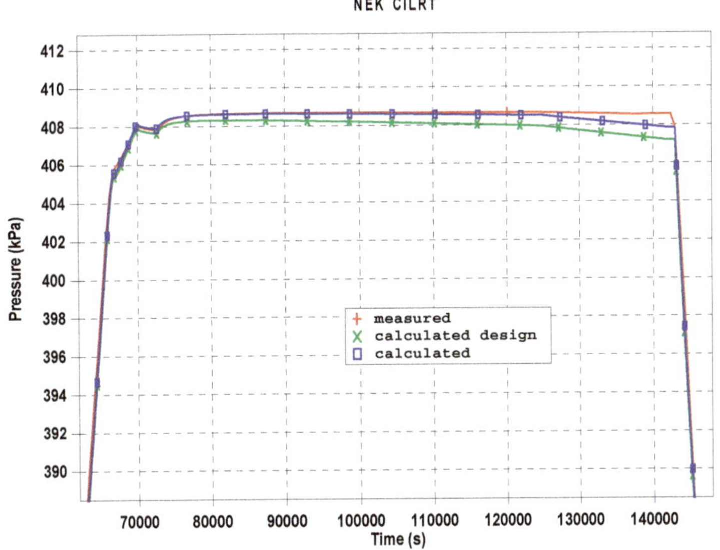

Figure 4. Decrease of the containment pressure due to leakage in the testing phase.

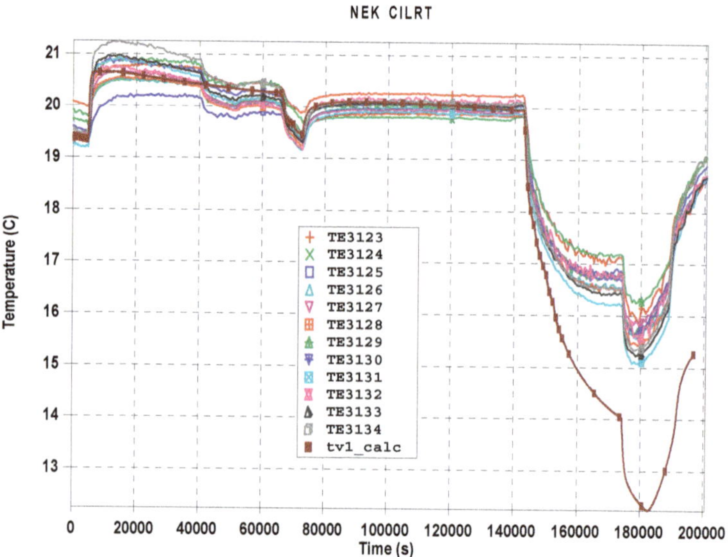

Figure 5. Measured and calculated containment temperatures during the test.

Measured (six positions) and calculated (RH1 calc, for RH in containment dome) relative humidity values are shown in Figure 6.

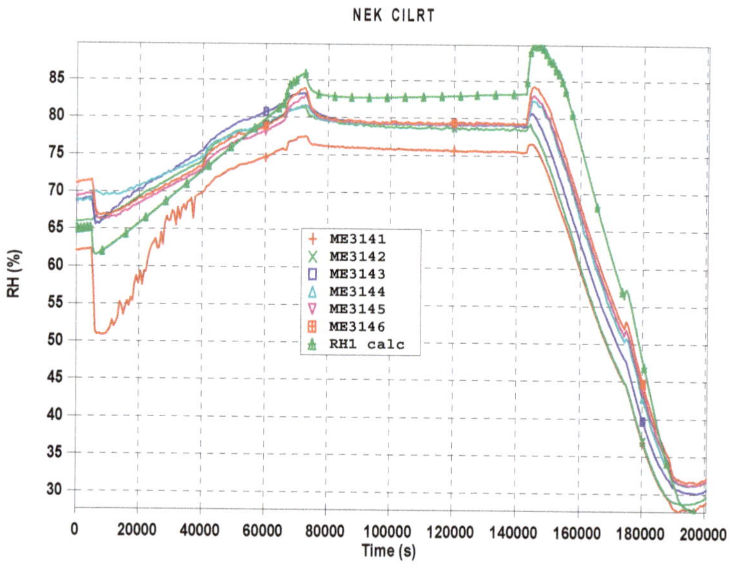

Figure 6. Measured and calculated containment relative humidity.

4.2. Sensitivity Calculations

Toward the end of the measuring phase, the increase of the air mass in the containment was calculated using measured data and the prescribed mass point method (Figure 7). The mass increase was characterized as an unphysical behavior because the high pressure in the containment was precluding any air inflow from the surrounding buildings or environment. The temperature transient alone can affect measurement results but cannot change amount of air available within the containment. In order to check reasons for such unphysical behavior, an additional run was performed by the GOTHIC code. A heat source with the power of 1 MW was introduced in the containment dome at 86400 s for the period of 100 s to determine sensitivity of the containment pressure and temperature (affecting measuring procedure) values on localized heat imbalance. In reality it was more likely that additional heat loss (instead of heat source) was present in the containment due to cold water in some systems.

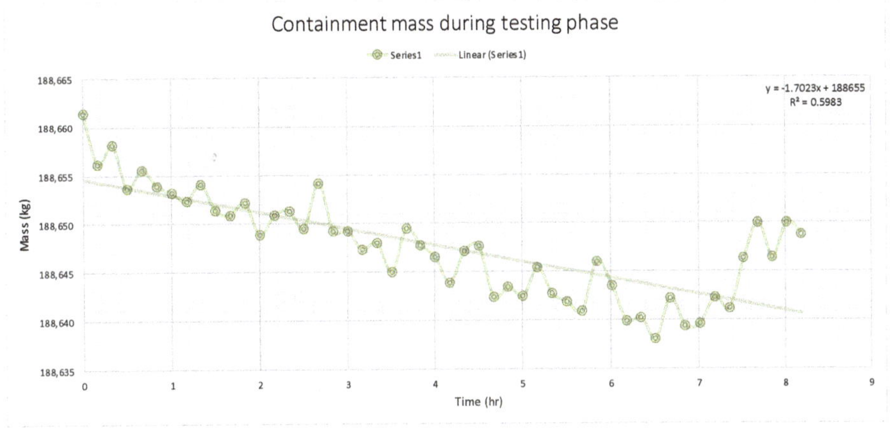

Figure 7. Calculated containment mass (linear fit) based on the testing phase measurement.

Sensitivity of the pressure to the presence of heat sources is shown in Figure 8. The resulting change in pressure and temperature distributions was not able to affect the calculation of air mass in the containment.

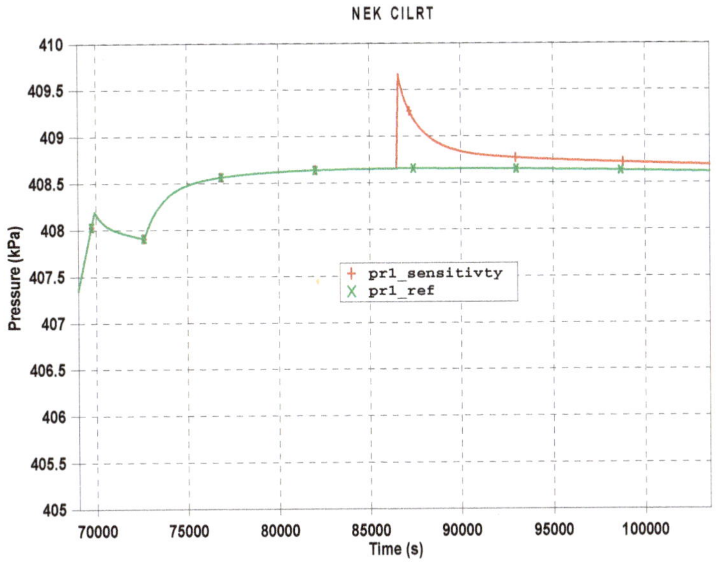

Figure 8. Containment pressure increase due to the addition of a localized heat source.

Another possible reason for such behavior could be related to the existence of air pockets having temperatures different than temperatures measured in available points and later redistribution of that air. The additional subdivision of the containment model could help to address that issue. We are planning to perform that calculation in next step. The main problem will be to find proper initial conditions (taking into account available measurements) and it will be more of the trial and error procedure to see what kind of initial distribution can cause a noticeable type of influence. In this paper, the primary interest was in collecting available measurements and performing the initial calculation. In addition, the model that was used in this article will be used for SAR calculations with real (based on CILRT) and design leakage. That way it will be possible to see the influence of real leakage on already performed calculations with conservative assumptions.

Taking into account that average values of temperature, RH and pressure were calculated from measured values using pre-calculated weighting factors we wanted to see what could be the influence of these factors to the calculated mass. The plant test procedure used volume factors to calculate volume averaged containment temperature from the temperatures measured at different locations. The temperature was used in the ideal gas equation to calculate dry air mass. The weighting factor is the ratio between the volume whose temperature is being "measured" by the detector located in the compartment and the total containment free volume. This paper used new factors calculated using a 3D model of the containment to post-process measured data (Figure 9). Even with an accurate 3D model, it was necessary to use some kind of arbitrary decision related to the placement of the boundary between more or less open areas.

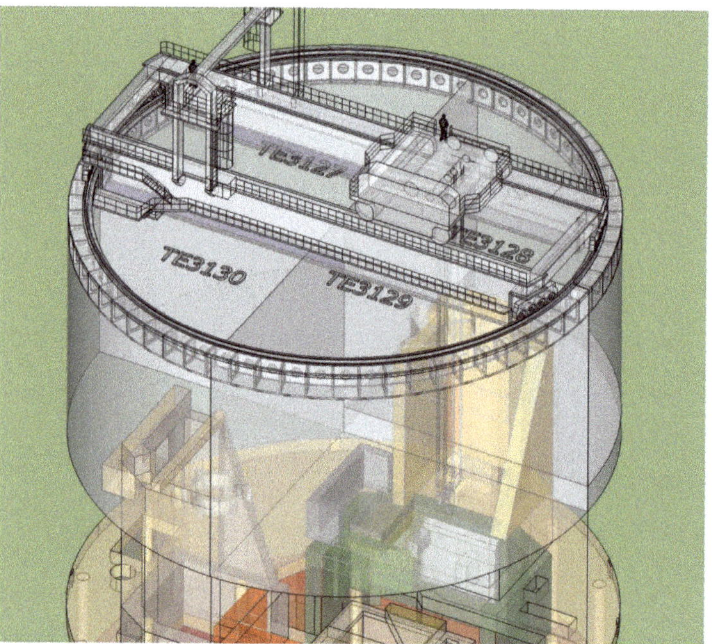

Figure 9. Subdivision of the upper containment part in the volumes for each instrumented location (volume weighting factors).

The old and new calculated volume weighting factors are shown for the available temperature measurement locations in Table 1 and for RH measurement locations in Table 2. The new weighting factors (even in the case when the change was rather significant), used with the measured temperature values, gave only a slightly higher average containment temperature (Figure 10) and, thus, lower containment mass, which means that the leakage rate was higher. Again, that influence could not explain the change in predicted air mass in the containment close to the end of measurement. The overall influence of new weighting factors to the calculation of leakage rates was relatively small ($L_{a,new}$ = 0.0217%/day, $L_{a,old}$ = 0.0207%/day), but they could be used to get more realistic leakage rates prediction in the future.

Table 1. Volume weighting factors for temperature measurements.

TE (Temperature Measurement Location)	Weighting Factors		TE (Temperature Measurement Location)	Weighting Factors	
	Old	New		Old	New
TE 3111	0.03	0.014106	TE 3123	0.068	0.046251
TE 3112	0.013	0.005118	TE 3124	0.058	0.057288
TE 3113	0.001	0.019656	TE 3125	0.049	0.042773
TE 3114	0.013	0.004162	TE 3126	0.053	0.059415
TE 3115	0.025	0.033087	TE 3127	0.073	0.077659
TE 3116	0.034	0.035967	TE 3128	0.073	0.075209
TE 3117	0.025	0.034235	TE 3129	0.073	0.077659
TE 3118	0.041	0.039981	TE 3130	0.073	0.075209
TE 3119	0.015	0.014223	TE 3131	0.058	0.058718
TE 3120	0.006	0.009623	TE 3132	0.058	0.058718
TE 3121	0.011	0.018807	TE 3133	0.058	0.058718
TE 3122	0.034	0.024691	TE 3134	0.058	0.058718

Table 2. Volume weighting factors for RH measurements.

ME (RH Measurement Location)	Weighting Factors	
	Old	New
ME 3141	0.057	0.043042
ME 3142	0.125	0.143270
ME 3143	0.066	0.067345
ME 3144	0.228	0.205730
ME 3145	0.292	0.305739
ME 3146	0.232	0.234875

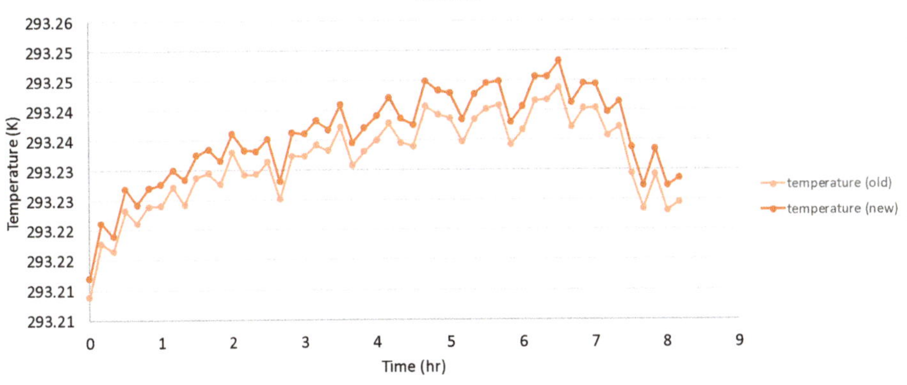

Figure 10. Average containment temperature change due to change in weighting factors.

5. Conclusions

During October 2016 an integrated leakage rate test was performed at full design pressure in the NPP Krško. Prerequisites for the test were local leakage rate tests of types B and C. The pressure, temperature, relative humidity, dew point temperature and the volumetric flow rate were measured during the test. The processing of measured data was carried out by the plant, according to the ANSI/ANS-56.8-2002 standard, using three standard procedures for calculating the containment leakage rate. The test results met all the criteria and showed that the leakage rate of the containment paths was below the prescribed limits.

In parallel to the processing of measured data, the calculation of the containment behavior was calculated using the GOTHIC 7.2b software package. The existing GOTHIC containment model was modified to reproduce the test results. Some of the data required for setting the model were not available and an iterative approach was used to determine the leakage flow path cross section. This way all the future analysis using this model can, in addition to the DBA leakage, take into account the proper (real) leakage rate of the system. It is now possible, during any possible accident, to see the pressure behavior of the containment and to estimate radiological consequences using a realistic leakage of the containment. To perform this type of the calculation, which is completely within the scope of the code, is just a matter of performing modeling in the proper way with proper data. It is expected that this type of calculation could be performed for any PWR (Pressurized Water Reactor) plant and similar agreement with the measured data could be obtained (if the required data is available and the test is properly conducted).

Toward the end of the measuring phase, the increase of the air mass in the containment was calculated using measured data and the prescribed mass point method. In order to check the reasons for such unphysical behavior, a sensitivity run was performed with the GOTHIC code and an additional heat source in the containment. We have expected that this run, together with careful examination of change in the containment and boundary conditions could help to find/quantify possible reasons for unphysical containment air mass increase toward the end of the test. The conclusion was that the measurement procedure is not inherently sensitive to localized change of the heat rate in containment. In addition, new weighting factors of the temperature and relative humidity sensors were determined using containment 3D geometry, just to be sure that representative average temperature and RH were correctly calculated from measured values. The influence of these new weighting factors is relatively small and unable to explain a change in the predicted air mass in the late phase of the test, but they can be used to get more realistic leakage rates in the future. In the next step we will try to explore, using a more detailed containment model, the possible influence of a single RCFC train operation in the initial phase of testing (asymmetry, air pockets) to test results.

Author Contributions: Conceptualization, S.V. and D.G.; methodology, S.V. and S.Š.; software, D.G.; validation, S.V., T.F. and V.B.; formal analysis, S.V. and S.Š.; investigation, S.V. and S.Š.; data curation, V.B.; writing—original draft preparation, S.V. and S.Š.; writing—review and editing, D.G.; visualization, T.F. and D.G.; supervision, D.G.

Funding: This research received no external funding.

Acknowledgments: The authors express their gratitude to the NPP Krško for providing the CILRT test data for the purpose of conducting GOTHIC simulations presented in the paper.

Conflicts of Interest: The authors declare no conflict of interest.

References

1. *NEK—Technical Specifications*; NPP Krško: Krško, Slovenia, June 2016.
2. *Containment Leakage Rate Testing Program, NEK Administrative Procedures*; NPP Krško: Krško, Slovenia, August 2013.
3. *ANSI/ANS-56.8-2002, R2016: Containment System Leakage Testing Requirements, Current Standard, Revision of ANSI/ANS-56.8-1994*; American Nuclear Society: La Grange Park, IL, USA, 2002.

4. Fancev, T.; Grgić, D.; Šadek, S. Verification of GOTHIC Multivolume Containment Model during NPP Krško DBA LOCA. In Proceedings of the 11th International Conference of the Croatian Nuclear Society, Zadar, Croatia, 5–8 June 2016.
5. *NAI 8907-06 Rev 17, GOTHIC Containment Analysis Package, Version 7.2b(QA)*; Technical Manual; EPRI: Palo Alto, CA, USA, March 2009.
6. *NAI 8907-02 Rev 18, GOTHIC Containment Analysis Package, Version 7.2b(QA)*; User's Manual; EPRI: Palo Alto, CA, USA, March 2009.
7. *NEK USAR*; NPP Krško: Krško, Slovenia, 2016; Chapter 6.

Article

Multiple-Criteria Decision-Making for Assessing the Enhanced Geothermal Systems

Sara Raos [1,*], Perica Ilak [1], Ivan Rajšl [1], Tena Bilić [1] and Ghislain Trullenque [2]

1 Faculty of Electrical Engineering and Computing, University of Zagreb, Unska 3, 10000 Zagreb, Croatia; perica.ilak@fer.hr (P.I.); ivan.rajsl@fer.hr (I.R.); tena.bilic@fer.hr (T.B.)

2 Département GEOS UniLaSalle, Équipe B2R, 19 rue Pierre Waguet, 60000 Beauvais, France; ghislain.trullenque@unilasalle.fr

* Correspondence: sara.raos@fer.hr

Received: 1 April 2019; Accepted: 24 April 2019; Published: 26 April 2019

Abstract: This paper presents the main features of a multiple-criteria decision-making tool for economic and environmental assessment of enhanced geothermal systems projects. The presented holistic approach takes into account important influencing factors such as technical specifications, geological characteristics, spatial data, energy and heat prices, and social and environmental impact. The multiple-criteria decision-making approach uses a weighted decision matrix for evaluating different enhanced geothermal systems alternatives based on a set of criterions which are defined and presented in this paper. The paper, defines and quantifies new criterions for assessing enhanced geothermal systems for a particular site. For evaluation of the relative importance of each criterion in decision making, the weight is associated with each of the listed criterions. The different scenarios of end-use applications are tested in the case study. Finally, in the case study, the data and statistics are collected from real geothermal plants. The case study provides results for several scenarios and the sensitivity analysis based on which the approach is validated. The proposed method is expected to be of great interest to investors and decision makers as it enables better risk mitigation.

Keywords: enhanced geothermal systems; economic assessment; environmental assessment; optimization; levelized cost of energy

---

1. Introduction

The growing concern around rising energy costs, the dependence on fossil fuels, and the environmental impact of energy supply makes it necessary to find economical and environment-friendly energy alternatives. The largest share in newly-installed power capacities around the world is covered by wind and solar plants. Besides those two renewables, geothermal energy also represents large untapped renewable potential and low environmental impact, especially regarding greenhouse gases emissions. Despite many other advantages, like a reliable, constant baseload electricity or direct heat usage and a small land area footprint, geothermal energy is nowadays still a small contributor to the primary energy consumption. Its worldwide installed capacity is estimated at 12.9 GW [1] and share in total electricity generation of less than 1%. The main reasons are related to the risks and uncertainties of sustained fluid provision from the reservoirs and large upfront costs associated with exploration, well drilling and stimulation [2]. Furthermore, the traditional hydrothermal systems, based on mature and well-known technology, enable the exploitation of mainly high-enthalpy reservoirs, whereas a huge geothermal potential is present in low permeable, low porosity and low to medium enthalpy bedrock. In order to enhance reservoir productivity in low permeable rocks, Enhanced Geothermal Systems (EGS) technology has been developed. The EGS technique consists of creating a fracture system in the targeted geological formation through which geothermal fluid can circulate. Hydraulic and chemical stimulations are used to create these fractures. The thermal energy stored in the hot rock mass gets extracted by circulating the fluid through the reservoir. The circulation of the fluid is obtained with different extraction technologies, among which the most prominent are so-called injector-producer doublets or borehole heat exchangers.

Aside from exploiting the geothermal heat from low permeable bedrock, additional energy can be recovered from thousands of mature or abandoned oil wells. Mature oil fields have been used for production for a long time, but their production has reached its peak and has started to decline. The typical production pattern of most oil producing wells displays an increase of water with time, from 0% initially to a point, typically above 95%, when it is not economic to produce the remaining oil. Mature oil fields account for more than 70% of world's oil and gas production, and regarding European oil fields, it is expected that wells are nowadays producing much more water than oil, with average water to a fluid ratio of 90% and with temperature up to 90 °C and sometimes higher. This remaining heat is currently usually wasted, as it is simply re-injected into the reservoir for pressure maintenance or sweep purpose. There are several studies concentrating on energy recovery from mature or abandoned oil fields. In Reference [3] a preliminary assessment of the potential for geothermal exploitation of the co-produced water from wells in the Villafortuna-Trecate oil field in Italy was made by comparing three different implementation scenarios for the possible use of the co-produced hot water: direct use district heating (DH), electric power generation through Organic Rankine Cycle (ORC) plant, and co-generation of heat and power. In Reference [4] energy from abandoned oil and gas reservoirs is used by oxidizing the residual oil with the injected air. In Reference [5] a simulation for the determination of geothermal power production from abandoned oil wells by injecting and retrieving a secondary fluid is performed. Technical feasibility study of acquiring geothermal energy from existing abandoned oil and gas wells is conducted in Reference [6]. Computational results indicate that the geothermal energy produced from abandoned wells depends largely on the flow rate of the fluid and the geothermal gradient. Moreover, the results also indicate that the distance of the two proposed wells should not be less than 40 m to avoid their interrelationships. A Poland case study on the usage of abandoned oil and gas wells for recovering geothermal heat is given in Reference [7]. Some important aspects of power generation using the co-produced hot oil and liquid, with temperatures around 120 °C, from Huabei oil field were studied in Reference [8]. Study on the geothermal power generation using abandoned oil wells is done in Reference [9] with isobutane as working fluid. Also, some interesting insights are given regarding how to increase obtained heat. The results in Reference [10] show that it is necessary to consider the oil and gas saturation while estimating geothermal reserve in oil and gas reservoirs. In Reference [11] the Croatian case of the binary power plant installed in Velika Ciglena is described and in Reference [12] the economic feasibility of this power plant is presented. Moreover, the thermodynamic cycle optimization of Velika Ciglena power plant is conducted in Reference [13]. The explorations in Croatia for the national oil company showed a very high temperature of geothermal water in the oil negative well (about 170 °C). Nowadays, the power plant at Velika Ciglena is the largest ORC geothermal power plant in Europe, with design conditions that allow for a 15 MWe installed capacity. The power plant is ORC using the isopentane as a working fluid and an air-cooling system that was chosen for the condensing process. Currently, the plant is producing electricity but there is still enough potential capacity for a DH, for which studies are currently being carried out. Furthermore, in Reference [14] a global review of 18 significant EGS sites and technologies that have been applied in the EU, Japan, South Korea, Australia, and the USA was given. The results from this study show that the site characteristics are a key factor of successful EGS development. It was concluded that for sites suitable for EGS, the local geological conditions (stress field, temperature field, rock composition, the range of existing permeability, reservoir properties, etc.) mostly determining the amount of recovered geothermal heat by means of EGS techniques.

Many previous studies have focused either on economic assessment [15,16] or environmental assessment based on the life cycle environmental impact of geothermal power generation as studied in References [17,18]. A review in Reference [19] presents an analysis of existing software packages for estimating and simulating costs, conventionally used in studying EGS facilities. The focus of the review is the top European software EURONAUT and the US GEOPHIRES package. EURONAUT is implemented based on the studies conducted at the EGS plant in Soultz-sous-Forêts. The root of the program is economic estimation via discontinuous cash flows, and all other calculations are developed as separate modules that can be joined together via various interfaces [19]. GEOPHIRES, however, is a software tool that combines reservoir, wellbore, and power plant models with capital and operating costs, correlations, and financial levelized cost model to assess the technical and economic performance of EGS. This software differs from the Geothermal Energy Technology Evaluation Model (GETEM) [20] and Hot Dry Rock economic model (HDRec) [21], which are two examples of technological/economic models initially used to simulate the operation of EGS plant. The distinction is mainly due to the fact that the currently available packages do not permit the simulation of not only electricity production but also direct-use-heat production and a combined heat and power (CHP) production, which was implemented in GEOPHIRES software.

The rate of geothermal development and implementation has been conditioned not only by mentioned geological conditions, drilling, and stimulation technologies, but also by legal frameworks, and regulative and social constraints. An overview of legislative and socio-economic issues of geothermal energy is available in References [22,23]. More on public and political acceptance issues could be found in Reference [24]. Establishing functional legal frameworks remains a challenge for countries seeking to develop their first geothermal projects to this day.

EGS technology enables exploitation of geothermal energy at a wide range of temperature and on a large geographic scale. However, EGS technology is not yet mature enough to be commercially competitive with other renewable resources. Almost all the EGS pilot plants, currently operating, need to be jointly funded by governments in order to operate or/and develop. Finally, the risks, uncertainties, and costs related to EGS projects mean that it is essential to conduct exhaustive studies involving modelling and simulation of EGS geothermal reservoirs and above-ground power plant facility at any location where it is desired to develop this kind of technology. This requires a holistic approach which should consider different scenarios and various influencing factors, from choosing the right extraction technology to the analysis of the energy prices and market signals. Given the fact that a taken decision will trigger financial consequences over a long period of time, the software for estimating and simulating the costs is an essential tool in order to successfully choose and face an EGS project. This paper presents the concept of such multi-scale Decision-Making Support Tool for Optimal Usage of Geothermal Energy (DMS-TOUGE) and multiple-criteria decision-making (MCDM) matrix that will be fully developed as part of the Horizon 2020 project: Multidisciplinary and multi-context demonstration of EGS exploration and Exploitation Techniques and potentials (MEET, GA No 792037).

The main objective of this work is to present the MCDM matrix and selected weighted criteria used to evaluate and compare different EGS sites and technologies and to demonstrate this MCDM matrix as a means of conducting a rapid preliminary evaluation of the technical and economic feasibility of an EGS project with related environmental and social impact. The MCDM matrix is used in this work for assessment of four different geothermal sites, and by using real data for each of the selected sites the comparison between sites was enabled.

The contribution of this work in relation to other selection matrices is that it provides expanded and detailed criteria related to environmental and social impact giving the necessary emphasis on so far neglected important aspects for successful completion of geothermal projects. Moreover, the novelty of this work is that by combining technical, economic, environmental, and social aspects of geothermal projects, the MCDM presented in this work gives a comprehensive assessment of EGS projects. Namely, since the EGS related projects are high-insensitive investments, for a DM interested in sustainably operate an EGS-plant close to densely populated areas, it is imperative to include many different aspects in techno-economic analysis and decision-making process. It is, however, challenging to predict all possible scenarios and influencing factors. Therefore, the process of developing a DMS-TOUGE and related MCDM matrix is a multiple-stage process. At this stage of the development and modelling process, the tool is intended for an evaluation of an EGS project at early development stage, meaning that a great number of the variables should be approximatively forecasted and estimated. Using optimization, the adequate evaluation of the production for chosen site and technology can be displayed. With progress in time, the tool will be useful for upgrade or extension of an already existing chosen geothermal site. In that case, the DM can provide more detailed information about EGS power plant operating cycle since the plant has already been generating electricity or heat for direct usage.

The rest of paper is organised as follows. Methodology, background, and main components of DMS-TOUGE are explained in Section 2. In the same chapter one of the main component of DMS-TOUGE, the MCDM is presented in more detail. Section 3 describes in detail the MCDM matrix and its criterions used in this paper for the evaluation of different EGS sites. Section 4 describes the case study examples, presents the scenario results, and gives the sensitivity analysis based on these results. The obtained results are discussed. Section 5 concludes the paper and gives the main directions for future development and research.

2. Background

2.1. Decision-Making Support Tool for Optimal Usage of Geothermal Energy (DMS-TOUGE)

The DMS-TOUGE represents a holistic approach for economic and environmental assessment of EGS sites that provides the capability of simultaneous site-specific environmental and economic analysis, among others considering low-enthalpy energy from co-produced hot water during oil fields exploitation. It considers existing infrastructure with possible extension or upgrades, the possible costs of future facilities, co-use/re-use of existing mature or abandoned oil fields boreholes, and other different geological contexts. Furthermore, the DMS-TOUGE will be useful for the decision makers (DM) involved in activities associated with: applications of EGS techniques to currently unexploited reservoir types (such as Variscan orogenic belt) by means of injector-producer doublets, increasing the productivity of existing power plants by reinjection of geothermal fluids with a colder temperature combined with the generation from the small-scale ORC units, and usage of hot fluids from mature and abandoned oil fields for electricity or heat production.

The DMS-TOUGE tool will use both external data entered directly by DM and internal data from database, depending on the quantity and quality of input data entered by a DM, such as: water temperature, geothermal capacity, electricity and heat prices, injection water flow rate values, turbine technology specifications, generator type, heat exchangers, working fluid, type of extraction technology (injector-producer doublets, deep borehole heat exchanger), risk analysis (thermal effect, possible appearance of scaling, radioactive deposits, mechanical evolution of casing) or environmental data (CO_2 emission schemes, security of energy supply issues). In order to better anticipate and include future events, different possible scenarios are evaluated and accounted and accordingly the tool can use forecasted data (load, prices, etc.) which can occur over the operation lifetime of EGS technology. Some of the most important environmental (external) factors that are integrated into the tool are: proximity of nearest suitable power system grid and/or nearest suitable district heating system where EGS could be integrated into, proximity and availability of water for water cooling mechanism, possibility for different usage of geothermal energy for agricultural, industrial and district heating needs, and geopolitical environment and relevant legislative framework. The DMS-TOUGE relies on optimization algorithms to value/quantify different technologies. It will be used to quantify environmental and social impacts and calculate system levelized cost of energy (sLCOE) of technology in order to find the best-suited option for a given site. Moreover, when assessing the best-suited option and technology for a given site, malicious and faulty components should be taken into account since, in reality, it is illusory to expect that all components of the geothermal system are functioning without problems. An example of addressing this issue is presented in Reference [25]. Therefore, it is mandatory to include any possible risks involved with different components of the system. As an integral part of the DMS-TOUGE, any possible risks will be analyzed, such as market (price) risks and technical issues:

thermal effects, scaling effect, radioactive deposits, and mechanical effect of the casing through the conditional value-at-risk (CVaR) measure [26–28]. Technical issues and risk of escalating costs need to be managed and hedged because of the potential damages likely to take place on an operational plant. The DMS-TOUGE will be verified and validated based on the comparison between tool output and real-life expert analyses on existing operating EGS sites and also historical data related to existing EGS sites. Output data of DMS-TOUGE will be available as raw data or in a form of directives and suggestions that are suitable for decision makers and investors. The raw data will be processed by a special subprocess, a separate MCDM, into a decision. The MCDM will be further discussed in Section 3. The schematic depiction of the main features of the DMS-TOUGE is shown in Figure 1. It should be emphasized that DMS-TOUGE as a whole is still in the development phase, and some parts have been modelled, including MCDM described later.

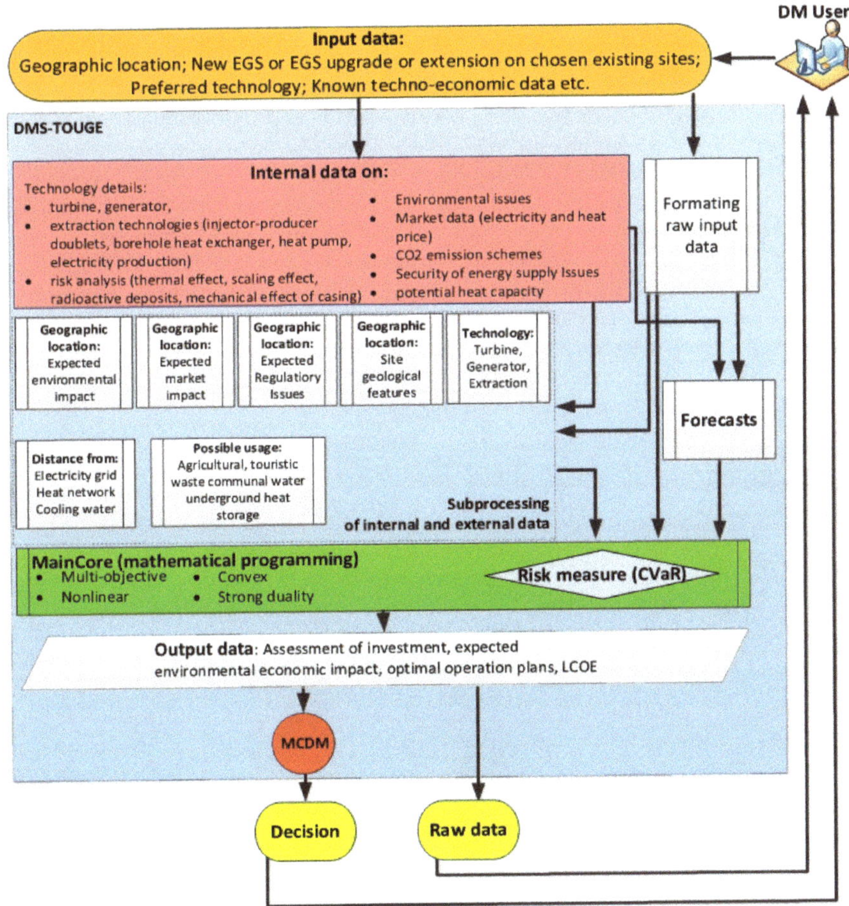

Figure 1. Schematic description of the main processes in Decision-Making Support Tool for Optimal Usage of Geothermal Energy (DMS-TOUGE).

2.2. Geothermal Project Development Phases

The geothermal project development occurs in various phases, with a unique duration of each phase. There is a series of seven key development steps before the actual operation and maintenance phase commences: (i) a preliminary survey, (ii) exploration, (iii) test drilling, (iv) project review and planning,

(v) field development, (vi) construction, and (vii) start-up and commissioning. The development of a typical utility size geothermal project will usually take between 5 to 10 years [29], depending on the country's geological conditions, information available about the resource, institutional and regulatory climate, access to suitable financing, and other factors. All seven-key development steps could be compressed in the first three following phases, followed by the fourth operation and maintenance (O&M) phase:

1. Discover and establish a viable resource;
2. Develop the project to the point necessary to obtain the power purchase agreement (PPA);
3. Complete the project development once the PPA is obtained, and
4. Operate the power plant facility.

Each phase includes activities and elements that incur different related costs, which in DMS-TOUGE can be estimated or inputted by a decision maker. These costs, along with the estimated power generation over a project's lifetime, are the basis for the sLCOE calculation for a defined scenario. The number of the activities in each phase, and consequently the capital and O&M costs of the project, depend also on the state of the project, i.e., is it a greenfield project representing a new EGS site or a brownfield project representing an existing EGS site with possible upgrade, or co-production and/or conversion of mature or abandoned oil fields. DMS-TOUGE will be able to calculate necessary costs for sLCOE estimation according to the input data regarding the above-mentioned steps and phases. Capital costs are present in the first three phases, and O&M cost in the fourth phase. All possibly occurring capital costs that will be estimated and calculated with DMS-TOUGE for defined EGS project are summarized in Figure 2. Furthermore, for brownfield project with existing infrastructure the drilling part is omitted and thereby approximately 40–60% of capital expenditure (CAPEX) is avoided, meaning that all the activities related to drilling process depicted in Figure 2. are omitted in that case.

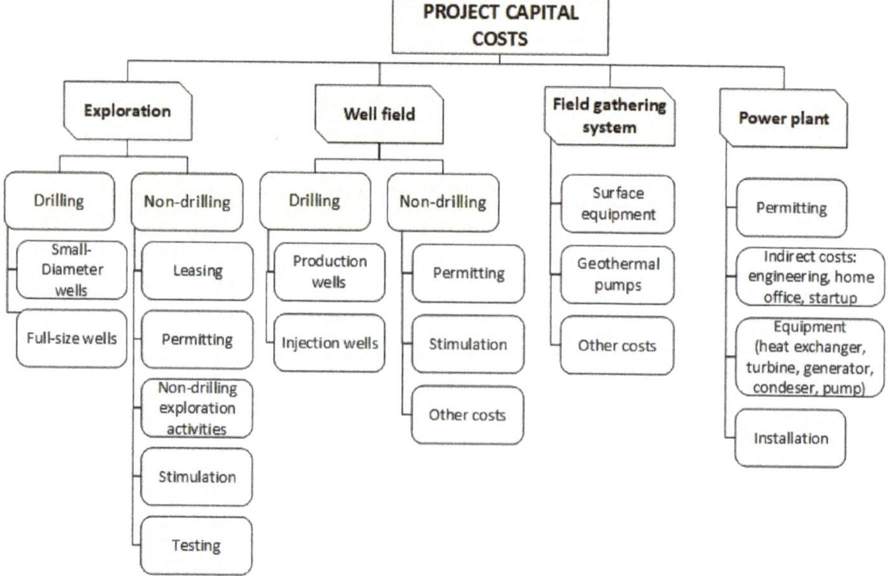

Figure 2. Capital costs included in Decision-Making Support Tool for Optimal Usage of Geothermal Energy (DMS-TOUGE) determination of a system levelized cost of energy (sLCOE).

2.3. Extraction Technology

Different extraction technologies need to be compared when assessing EGS sites, such as the producer-injector doublet and deep borehole heat exchanger. The production system, which consists of a production well, a surface heat exchanger, and a reinjection well, with all produced and cooled thermal water, returned to the aquifer is the so-called doublet system (Figure 3a). The production and the reinjection well must be drilled to the same geological horizon, i.e., reservoir. Consequently, the geofluid circulates in a closed loop: from the reservoir through the production well to the heat exchanger and back to the reservoir through reinjection wells. It is worth noting that pumping the fluid back in the reservoir maintains its hydrostatic pressure. (In the reservoir, around the bottom of the injection well, the permanently inflowing cooled water forms a cold front which propagates toward the production well). Borehole heat exchanger for shallower depths consists of single (Figure 3b) or double U-tubes (Figure 3c). In deeper wellbores, the presence of a liner will entail a coaxial-type solution (Figure 3d), which better optimises thermal and hydraulic behaviour. Injection takes place through the annulus and production (or heat delivery) through a centred co-axial pipe.

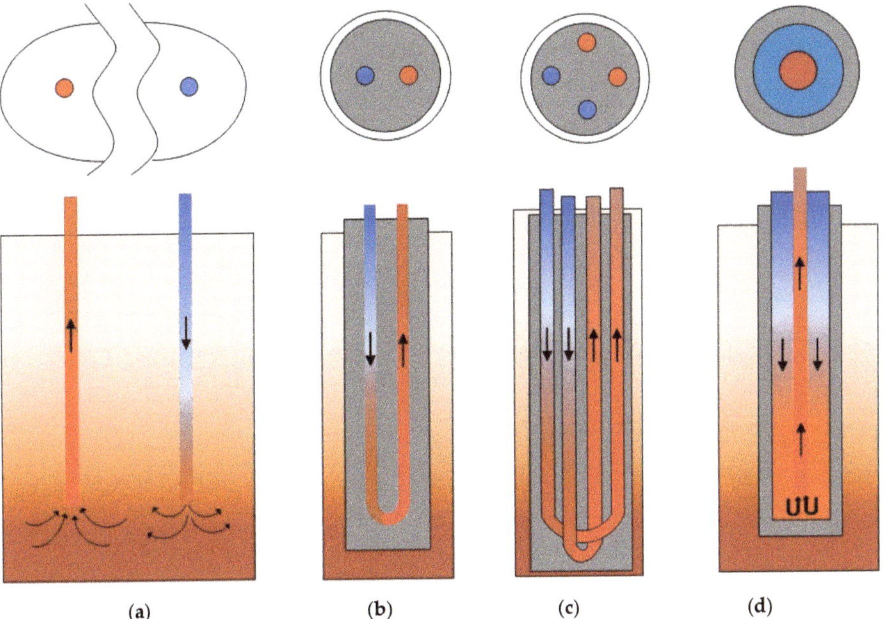

Figure 3. Extraction technologies: (**a**) producer-injector doublet, (**b**) borehole heat exchanger single U-tube, (**c**) borehole heat exchanger double U-tube, (**d**) coaxial borehole heat exchanger.

2.4. Geothermal Facility Arrangement

Different end-user options need to be compared with the simulation and optimization of three different end-user scenarios; (i) only direct-use heat production; (ii) only electricity production, and (iii) CHP production. The geothermal heat can be delivered to a nearby DH network and/or power plant via heat exchangers. Figure 4 presents a schematic description of possible end-user options. In the case of direct use option, (Figure 4a) the thermal energy from the geothermal fluid is transferred to the working fluid, usually water or steam, in DH network via heat exchanger. In the case of electricity production, the thermal energy is transferred to the working fluid circulating in the secondary system, ORC, also via heat exchanger, as depicted in Figure 4b. For the third end-user option (CHP) two configurations of the CHP plant are proposed: series and parallel, respectively. For the series configuration (Figure 4c) the brine (geothermal water) delivers the heat to the ORC at a higher temperature and subsequently to the DH system or other direct use application at a lower temperature. The flow rate of the brine is constant and equal for both systems. In the case of the parallel configuration (Figure 4d), geothermal heat is delivered to both the DH system and the ORC at a high temperature but with lower flow rate compared to a series configuration. The selection of the configuration depends on the features of the selected geothermal site and its surrounding, as well as on heat characteristics needed in the connected DH system.

A default power plant type is based on the resource temperature [30] and since the focus of the research are low enthalpy resources, i.e., medium and low-temperature geothermal fields with temperatures of 200 °C and lower, the binary power plants show good performance and therefore will be modelled in DMS-TOUGE. The concept of the binary power plant is based on the ORC. In the ORC system, the fluid flow is divided into four steps. The working fluid is heated up and vaporized in a hot heat exchanger (the evaporator), and at this point, the temperature of the working fluid is the highest. This saturated steam drives the turbine, which enables electricity production by lowering the fluid pressure to its low level. The fluid is then condensed in a cold heat exchanger (condenser) and pumped again into the evaporator, increasing thereby the fluid's pressure. As for the condensers, the most commonly used heat rejection equipment at binary geothermal power plants are the air-cooled condensers. The use of air eliminates the requirement for makeup water. However, compared to water cooling towers, air-cooled condensers require more space and represent a larger parasitic power load on the plant.

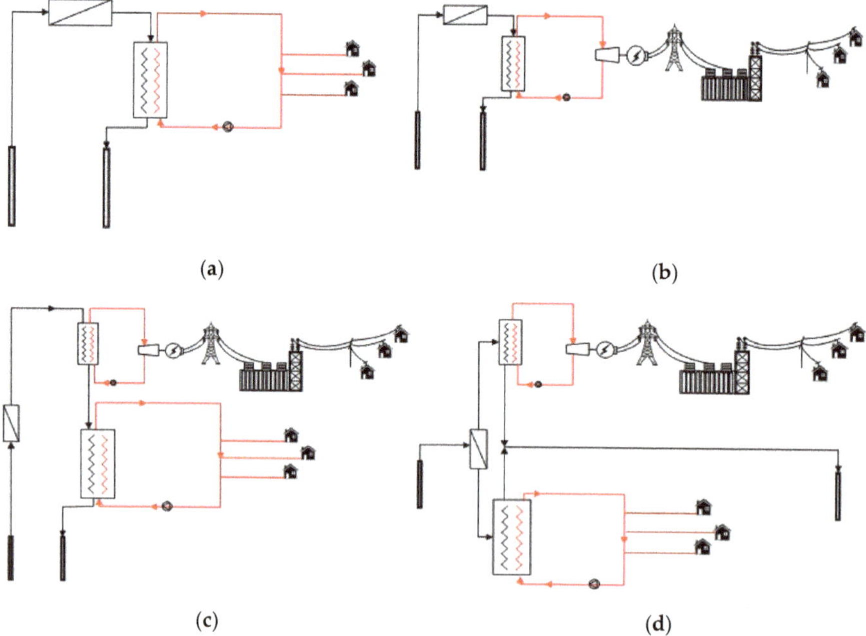

Figure 4. Schematic description of the possible end-user options: (**a**) direct use, (**b**) only electricity production, (**c**) combined heat and power (CHP)–series configuration, (**d**) CHP–parallel configuration.

The geothermal fluid, the hot source of the heat, is circulated in another loop on the other end of the evaporator. Therefore, for the binary plants, the major equipment components estimated are turbine-generator, air-cooled condenser, geothermal heat exchangers, and working fluid pumps, as depicted in Figure 5.

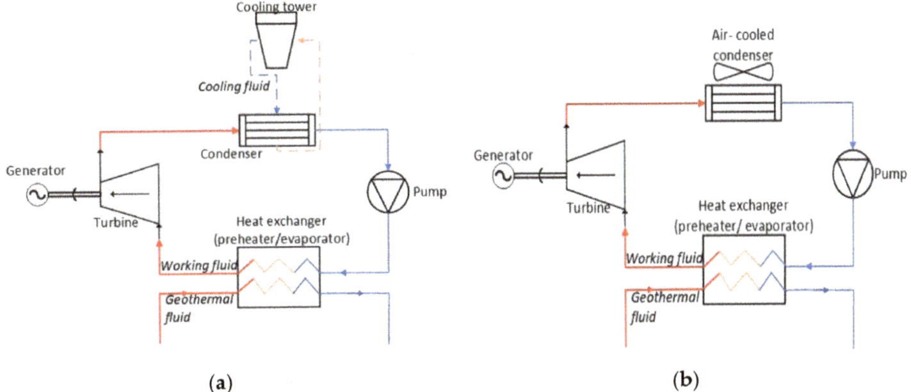

Figure 5. Schematic diagram of a basic binary cycle geothermal power plant: (**a**) water-cooled condenser and (**b**) air-cooled condenser.

3. Multiple-Criteria Decision-Making Analysis

As a subprocess in the DMS-TOUGE, an MCDM analysis will be used and performed using the weighted decision matrix (WDM). For evaluating different EGS options, a set of criterions is defined and presented in this subsection. For evaluation of the relative importance of each criterion in decision making, the weight is associated with each of the listed criterions. Performance, x_{ij}, of option i on criterion j is arbitrarily defined with a numerical value from 1 to 5, whose higher value means better performance, $x_{ij} \in \{1,2,3,4,5\}$. Finally, total performance, X_i, of ith EGS option on all criteria, $\forall j$, is assessed by summing all performance values, x_{ij}, multiplied by its weight as defined in Equation (1).

$$X_i = \sum\nolimits_j^J w_j \cdot x_{ij}, \tag{1}$$

where X_i is the total performance of ith EGS option, $i \in I$, where I is a total number of EGS options. The w_j is weight i.e., relative importance in the decision making of criterion j, $j \in J$, where J is a total number of criteria. The x_{ij} is the performance of option i on criterion j. To use WDM successfully while assessing EGS options for a specific geothermal site, a set of well-defined criteria is needed. The criteria on which EGS option will be evaluated is listed below in Sections 3.1–3.12. It should be noted that, although the Equation (1) shows a simple summation, the actual calculation of total performance for a selected site was performed as an average mean of X_i performances of each criterion. Also, in this work every criterion j, is assumed to have equal relative importance in the decision making. Thereby, all criteria are evaluated assuming all w_j to be equal and valued with 1. It is worth mentioning that the actual range for w_j has not been yet precisely defined, as it is a sensitive part of the MCDM matrix. Therefore, such a range must be chosen that will adequately and meaningfully reflect the relative importance of certain criteria in relation to others. In the future work and through a process of validation and verification of the MCDM matrix and DMS-TOUGE, the discrete sequential values for w_j will be chosen.

3.1. First Criterion: Installed Power, $x_{i,1}$

When considering energy investment, installed power (capacity) is the first and most important parameter [31]. It later determines both costs (CAPEX, OPEX-O&M) and revenues (power output). According to Reference [30], performance $x_{i,1}$ of option i on 1st criterion should be determined by P/P_r ratio in p.u. (ratio of installed power, P, of ORC technology in ith option to the reference installed power, P_r, e.g., according to Reference [31] for ORC technology reference installed power ranges from 1 MW to 5 MW depending on site features) (Table 1).

Table 1. Performance values $x_{i,1}$ for 1st criterion.

Ratio (*p.u.*)	$0 \leq P/P_r < 0.3$	$0.3 \leq P/P_r < 0.6$	$0.6 \leq P/P_r < 0.9$	$0.9 \leq P/P_r < 1.2$	$1.2 \leq P/P_r < \infty$
$x_{i,1}$	1	2	3	4	5

3.2. Second Criterion: Fluid Heat Flow, $x_{i,2}$

Expected heat flow, Q (W), for the two main extraction technologies, a traditional doublet with fluid extraction and reinjection, and wellbore heat exchanger, a closed loop system, is defined with Equation (2):

$$Q = q \cdot \rho \cdot c_p \cdot (T_H - T_C), \tag{2}$$

where q is the fluid flow rate (m^3/s), ρ is fluid density (kg/m^3), c_p is specific heat capacity of fluid at constant pressure (J/kg·K). T_H is the fluid temperature at the wellhead (°C), T_C is the fluid temperature at the exit of the steam turbine (°C).

Heat flow as a criterion was first proposed in Reference [31] and is used here with modifications in means of corresponding ranges for temperature. The idea is to emphasize the importance of flow rate and temperature of the produced fluid, and the impact of selected technology on flow rate and temperature. Reference [31] proposes criterion, here referred to as the $x_{i,2}$, whose value is based on the ratio between the fluid flow rate, q, and flowing temperature at the wellhead, T_H. According to [31] best suited ranges for valuing performance $x_{i,2}$ in heat flow criterion are defined by flow rates between 0 m^3/h and 100 m^3/h and when temperatures are between 80 °C and 160 °C, since those ranges correspond to operative conditions for an ORC plant. For this criterion, the temperature range is modified for the values between 60 °C and 160 °C, since one of the focus of the ongoing H2020 MEET project is to make those temperatures also exploitable (Table 2).

Table 2. Performance values $x_{i,2}$ for 2nd criterion.

Ratio	$1.67 \leq \frac{q}{T_H} < \infty$	$0.679 \leq \frac{q}{T_H} < 1.67$	$0.357 \leq \frac{q}{T_H} < 0.679$	$0.056 \leq \frac{q}{T_H} < 0.357$	$0 \leq \frac{q}{T_H} < 0.056$
$x_{i,2}$	1	2	3	4	5

3.3. Third Criterion: Theoretical Maximum Efficiency, $x_{i,3}$

In this paper, the ORC based power plants are proposed for exclusively electricity generation. It is mainly due to the low-to-medium temperature range of the produced geothermal fluid. The thermal efficiency evaluated at the heat exchanger of the conversion plant in such fields is usually less than 10%, and for the binary power plants, according to Reference [32] could be calculated with Equation (3).

$$\eta_{max} = 6.9681 \cdot \ln(T_H) - 29.713, \tag{3}$$

where the T_H represents the temperature of the produced fluid at the wellhead in (°C) and η_{max} is expressed in (%).

However, in case of power plants different from ORC, meaning other types of conversion plants, the thermal exchanged cycle between the two fluids, one circulating in the primary cycle (geofluid) and other circulating in the secondary circle (working fluid), can be assessed using the Carnot's ideal efficiency. In those cases, the expected theoretical maximum efficiency of conversion, η_{max} expressed in (%), is evaluated using the Equation (4) as according to Reference [33].

$$\eta_{max} = (1 - T_C/T_H) \cdot 100\%, \tag{4}$$

and it depends on the geological site features (the T_H part, where T_H in (K) is the outlet temperature from the borehole at the wellhead), technology and environment features (the T_C part, where T_C in (K) is the inlet temperature at the wellhead). The performance is valued as shown in Table 3. Depending on the end-user option the Equations (3) or (4) are used for assessment of the selected site.

Table 3. Performance values $x_{i,3}$ for 3rd criterion.

Efficiency of Conversion for ORC Plant (%)	$\eta_{max} < 4$	$4 \leq \eta_{max} < 6$	$6 \leq \eta_{max} < 10$	$10 \leq \eta_{max} < 12$	$\eta_{max} \geq 12$
$x_{i,3}$	1	2	3	4	5
Efficiency of Conversion for Other Cases (%)	$\eta_{max} < 30$	$30 \leq \eta_{max} < 40$	$40 \leq \eta_{max} < 50$	$50 \leq \eta_{max} < 60$	$\eta_{max} \geq 60$
$x_{i,3}$	1	2	3	4	5

3.4. *Fourth Criterion: Geothermal Gradient, $x_{i,4}$*

When setting the starting point of the geothermal plant feasibility analysis the geological factors should be considered. The efficiency of the heat transfer through the wellbore is highly dependent on the reservoir's initial temperature, which is a function of the well depth. Also, high thermal conductivity is required, so that the heat stored in the rocks could be transferred to the wellbore fluid. According to Reference [34], these two influencing factors could be collectively combined and represented with geothermal gradient, G_T (°C/100 m). The paper also suggested a range of geothermal gradient based on measured gradients for several analyzed oil fields across the world, which was taken for evaluating the performance $x_{i,4}$ in the geothermal gradient criterion (Table 4).

Table 4. Performance values $x_{i,4}$ for 4th criterion.

Geo.gra. (°C/100 m)	$G_T < 0.5$	$0.5 \leq G_T < 2$	$2 \leq G_T < 4$	$4 \leq G_T < 6$	$G_T \geq 6$
$x_{i,4}$	1	2	3	4	5

3.5. *Fifth Criterion: The Fluid Temperature at Wellhead, $x_{i,5}$*

According to Reference [31] the outlet temperature of the fluid from the wellhead is one of the main features of the geological site. It later determines installed power, technology, efficiency, revenues, and costs. Namely, this criterion has been defined to emphasize the relevance of this temperature, since it has a great impact on the conversion cycle in the power plant. The higher the fluid temperature at the wellhead, the higher the amount of the heat that can potentially be transferred to the secondary working fluid (depending of course on the heat exchanger technology). Performance $x_{i,5}$ of option on 5th criterion increases linearly depending on fluid temperature, T_H (Table 5). Here, the focus is on the utilization of temperatures from 60 °C to 160 °C (although upper bound for analyzed geo sites in the scope of this research is expected to be 90 °C) for cases of smart ORC units. This approach is adapted and modified from the work [31].

Table 5. Performance values $x_{i,5}$ for 5th criterion.

Temp. (°C)	$T_H \leq 60$	$60 < T_H \leq 90$	$90 < T_H \leq 120$	$120 < T_H \leq 150$	$150 < T_H \leq \infty$
$x_{i,5}$	1	2	3	4	5

3.6. *Sixth Criterion: Global Efficiency, $x_{i,6}$*

Aside from the geological setting and wellbore conditions, the supply of heat or/and electricity is directly related to the performance of the plant energy conversion. Therefore, a global efficiency criterion should be established to evaluate the multi-stage heat (energy) loss within the energy conversion cycle and the impact of the ambient temperature on stored heat (Table 6). According to

Reference [33] the total heat loss is addressed by means of coefficients of the different stages of the conversion cycle resulting in an overall evaluation for plant conversion. Equations (5)–(7) defines those coefficients, where Equation (5) represents the heat loss due to Non-Condensable Gases (NCG), where C [%] is the estimates of NCG weight, because the presence of NCG can negatively impact the operation of the plant turbine. The Equation (6) represents the parasitic load heat loss, including well pumps, cooling tower, condenser, where P_{TPL} is total parasitic load and P_{gross} gross thermal power. Moreover, Equation (7) defines the parasitic loss during the working fluid transport where the L_p is the pipe length in km.

$$\eta_{NCG} = 1 - 0.0059 \cdot C \quad (5)$$

$$\eta_{TPL} = 1 - P_{TPL}/P_{gross} \quad (6)$$

$$\eta_{pipe} = 1 - 0.003 \cdot L_p \quad (7)$$

Overall plant efficiency is then calculated according to the Equations (8)–(10), where Equation (8) represents evaluation of the electricity generation, Equation (9) is used to evaluate combined heat-electricity production (CHP), where the second heat exchanger is required to exploit the remaining thermal energy of geothermal water into another district heating fluid, and Equation (10) represents the evaluation of direct usage of heat only in DH systems. Moreover, to measure the operational performance of the turbine and the generator efficiency, η_t and η_g, respectively, were included.

$$\eta_{G(E)} = \eta_{max} \cdot \eta_{NCG} \cdot \eta_t \cdot \eta_g \cdot \eta_{TPL} \cdot \eta_{pipe} \quad (8)$$

$$\eta_{G(CHP)} = \eta_{max1} \cdot \eta_{NCG} \cdot \eta_t \cdot \eta_g \cdot \eta_{TPL} \cdot \eta_{pipe} \cdot \eta_{max2} \quad (9)$$

$$\eta_{G(DH)} = \eta_{max} \cdot \eta_{pipe} \cdot \eta_{TPL} \quad (10)$$

where η_{max1} in the equation related to the CHP represents the efficiency of conversion in ORC for electricity production and η_{max2} is the efficiency of conversion of the remaining heat from the geothermal fluid for direct heat usage.

Table 6. Performance values $x_{i,6}$ for 6th criterion.

Global Efficiency	$\eta_G < 0.2$	$0.2 \leq \eta_G < 0.3$	$0.3 \leq \eta_G < 0.4$	$0.4 \leq \eta_G < 0.6$	$\eta_G \geq 0.6$
$x_{i,6}$	1	2	3	4	5

3.7. Seventh Criterion: Corrosion and Scaling Hazard, $x_{i,7}$

The corrosive or scaling tendency of the geothermal site is evaluated with the Langelier Saturation Index (LSI) (Table 7). The LSI later determines O&M costs. The less the LSI, the better the performance of the option will be (see Reference [31]).

Table 7. Performance values $x_{i,7}$ for 7th criterion.

LSI	1.5 < **LSI** ≤ 2	1 < **LSI** ≤ 1.5	0.5 < **LSI** ≤ 1	0 < **LSI** ≤ 0.5	**LSI** = 0
$x_{i,7}$	1	2	3	4	5

3.8. Eighth Criterion: Distance from Power/Heating Grid, $x_{i,8}$

Construction of power lines and substations for plant connection to the grid imposes significant costs and should be addressed accordingly. Therefore, the distance between the geothermal plant production site and the nearest power and/or district heating system connection point must be addressed. Depending on that distance, *d* (km), the investment costs and also sLCOE vary. Apart from the distance, considering that the grid connection costs are site-specific, there are many other factors that have an impact on the investment costs. Due to the complexity if all the factors were included, the main influencing factor, namely the distance, was taken for the evaluation of the performance $x_{i,8}$. The range is shown in Table 8, finishing with the most favored onsite use, i.e., a very small distance between the plant facility and the connection point.

Table 8. Performance values $x_{i,8}$ for 8th criterion.

Distance (km)	$d > 4$	$3 \leq d < 4$	$2 \leq d < 3$	$1 \leq d < 2$	$d < 1$
$x_{i,8}$	1	2	3	4	5

3.9. *Ninth Criterion: Load Factor,* $x_{i,9}$

Generally, combined heat and electricity production increases the net efficiency of the power plant, which in turn improves power plant economics. This is even more important in the case of geothermal plants, where thermodynamic efficiencies are typically much lower compared to conventional power plants, due to the lower working fluid temperatures. Considering direct-use systems, heat is only supplied to the process, like greenhouse or district heating, when it is needed. As a result, according to Reference [34], the load factor, f_L, can vary from 15% to 75% depending on the application. Reference [34] examined the costs of the delivered heat as a function of a load factor for U.S. climates. Knowing that industrial applications can have a load factor of 0.30 to 0.75, the space heating only application 0.15 to 0.30, aquaculture 0.50 to 0.80 and greenhouses 0.18 to 0.24, the results showed that the high load factor correlates with lower cost of delivered heat and consequently affects the project's economic feasibility and sLCOE. As for electricity production, the load factor is typically 0.8 and higher. For combined heat and electricity production, the load factor is higher than in the case of only heat production. Based on forenamed results, the range for performance $x_{i,9}$ of alternative is determined and showed in the Table 9.

Table 9. Performance values $x_{i,9}$ for 9th criterion.

Load Factor	$f_L \leq 0.2$	$0.2 < f_L \leq 0.4$	$0.4 < f_L \leq 0.6$	$0.6 \leq f_L \leq 0.8$	$0.8 < f_L \leq 1$
$x_{i,9}$	1	2	3	4	5

3.10. *Tenth Criterion: sLCOE,* $x_{i,10}$

The average cost of the project over the lifetime will be addressed by the sLCOE (system LCOE, depending on the end-user option) in (€C/MWh) which also accounts for the costs of integration. This sLCOE will be calculated from all the available external data provided from DM and/or internal data from the internal database of the DMS-TOUGE, and according to Equation (13). Performance $x_{i,10}$ of option i on sLCOE criterion is determined by $sLCOE/\overline{\pi}$ ratio in p.u. (ratio of sLCOE of ORC technology in ith option to the average market price, $\overline{\pi}$, in different forecasts and for different horizons) shown in Table 10.

Table 10. Performance values $x_{i,10}$ for 10th criterion.

Ratio	$1 \leq \frac{sLCOE}{\overline{\pi}} < \infty$	$0.8 \leq \frac{sLCOE}{\overline{\pi}} < 1$	$0.6 \leq \frac{sLCOE}{\overline{\pi}} < 0.8$	$0.4 \leq \frac{sLCOE}{\overline{\pi}} < 0.6$	$0 \leq \frac{sLCOE}{\overline{\pi}} < 0.4$
$x_{i,10}$	1	2	3	4	5

3.11. *Eleventh Criterion: Social Impact,* $x_{i,11}$

According to Reference [22], the social acceptability is conditioned by the deviation from the regular condition in the area and utility of the affected parties from the project. As geothermal technologies are site-specific (the geology is different all over Europe and knowledge of the local geological setting is essential) and capital-intensive, the needs regarding exploration, resource development, construction, and O&M are covered by the local workforce. According to Reference [35], the costs of social acceptance could be presented as the external costs of a geothermal project. Depending on the site, type, and size of the project the amount of those external costs range, on the average, 0.5–2% and 1.5–4% of the total construction costs, i.e., approximately 17,000–220,000 € and 265,000–7,040,000 €, for direct use and multi-purpose projects, respectively. Moreover, employment potential could be divided into direct, indirect, and induced employment effect and quantified in terms of full-time jobs/MW (FT) and person × years of construction and manufacturing employment (Table 11). Total direct, indirect, and induced employment ratio is a ratio of the installed capacity, $P_{inst.}$, (MW) and full-time jobs calculated from the Equation (11) (shown in Figure 6). Equation (12) represents construction and manufacturing employment (C&M), where those jobs are expressed as full-time positions over one year (person × year) as a function of installed capacity, $P_{inst.}$, (MW). However, those C&M jobs are spread over several years taking into account the development time frame for the new projects. The social impact criterion will be obtained by the average of performances of the following sub-criterions: $x_{i,11,1}$ social acceptance costs of direct use or electricity production sub-criterion, $x_{i,11,2}$ social acceptance costs of combined heat-electricity production sub-criterion, $x_{i,11,3}$ FT employment sub-criterion, $x_{i,11,4}$ C&M employment sub-criterion. Each sub-criterion will be evaluated with a weight in a range from 1 to 5. Figure 5 shows the full-time jobs function as a defined in Equation (11).

$$FT\ jobs = log_{1.068}(P_{inst.}) \tag{11}$$

$$C\&M\ jobs = 22.4 \cdot P_{inst.} \tag{12}$$

Table 11. Performance values $x_{i,11}$ for 11th criterion.

Social Acceptance Costs of Direct Use or Electricity Production $(\cdot 10^3)$	$sac_{DU} > 295$	$145 < sac_{DU} \leq 295$	$30 < sac_{DU} \leq 145$	$4.5 < sac_{DU} \leq 30$	$sac_{DU} \leq 4.5$
$x_{i,11,1}$	1	2	3	4	5
Social Acceptance Costs of Combined Heat – Electricity $(\cdot 10^3)$	$sac_{CHP} > 6155$	$2640 < sac_{CHP} \leq 6155$	$880 < sac_{CHP} \leq 2640$	$350 < sac_{CHP} \leq 880$	$sac_{CHP} \leq 350$
$x_{i,11,2}$	1	2	3	4	5
Employment FT $\left(\frac{dFT}{dP_{inst.}}\right)$	$e_{FT} < 1$	$1 \leq e_{FT} < 1.5$	$1.5 \leq e_{FT} < 2$	$2 \leq e_{FT} < 4$	$e_{FT} \geq 4$
$x_{i,11,3}$	1	2	3	4	5
Employment C&M $(person \times year)$	$e_{C\&M} \leq 50$	$50 < e_{C\&M} \leq 150$	$150 < e_{C\&M} \leq 250$	$250 < e_{C\&M} \leq 350$	$e_{C\&M} > 350$
$x_{i,11,4}$	1	2	3	4	5
Total Social Impact	$AV \leq 1$	$1 < AV \leq 2$	$2 < AV \leq 3$	$3 < AV \leq 4$	$4 < AV \leq 5$
$x_{i,11}$	1	2	3	4	5

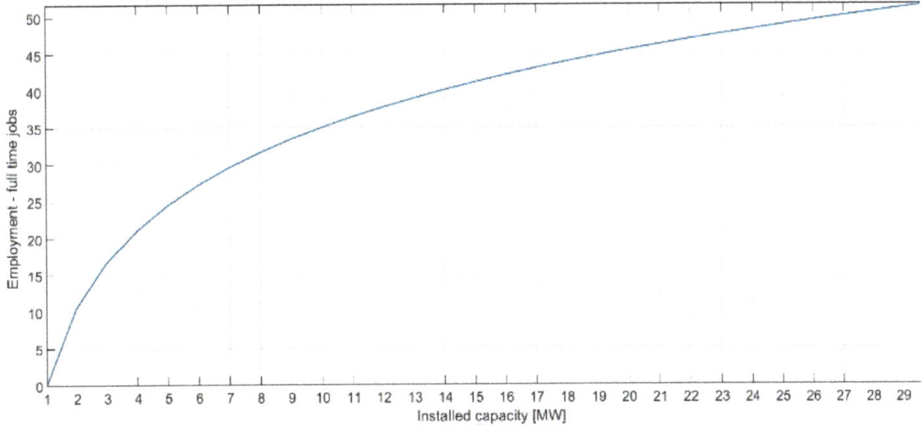

Figure 6. Total employment – full-time jobs function depending on the installed capacity of the project

3.12. *Twelfth Criterion: Environmental Impact,* $x_{i,12}$

According to Reference [24], the environmental impact should account for the impact on sustainability, landscape, subsidence, and potentially induced micro-seismicity, and account for the level of noise, atmospheric emissions, potential water contamination, and production of radioactive scaling. The fluid extraction could cause subsidence because of reservoir pressure decline and uplift during borehole drilling activities. This is measured in mm/year of soil decay. Moreover, the pore pressure reduction in production and increase in reinjection operations have been associated with increased induced seismicity, often microseisms of low energy (<2–3 M Richter scale) [31]. According to Reference [36] the ranges for this sub-criterion were assigned. The impact on the landscape is measured as land use intensity (LUI) for installed power in m^2/kW, and the range was estimated according to Reference [37]. The noise impact during routine operation is mainly caused by cooling towers and electrical transformers, and according to Reference [38], typical acceptable levels are 71–83 dB at 900 m distance from the facility. When considering atmospheric emissions, closed cycles, such as binary plants, have no gaseous emissions or they are close to zero and so do not contribute to air pollution. Considering objects of interest in this research (Figure 3), the impact on surface waters can be excluded. Groundwater contamination may occur if the casings in reinjection wells should fail, allowing fluid to leak. According to WHO, the range of total dissolved solids (TDS) and pH values was determined for the quantification of this sub-criterion. Radioactivity is mainly caused by interaction between the geothermal fluid and certain formations containing radioactive elements. As emphasized in Reference [39], the content of radionuclides in acidic magmatic rocks is generally higher compared to sedimentary rocks. Furthermore, uranium (U) and thorium (Th) are the most common radioactive elements found in granites. As defined in Reference [39], different types of rocks can contain different radioelements. Therefore, this sub-criterion is defined with the ranges of the forenamed most common radioactive elements but also taking into account the type of the rock – granitic rocks, shales, basaltic rocks, sandstones, and carbonates - in ascending order of the weights. The environmental impact criterion will be obtained by the average of performances of the following sub-criterions shown in Table 12: $x_{i,12,1}$ subsidence sub-criterion, $x_{i,12,2}$ potential seismicity sub-criterion, $x_{i,12,3}$ land use sub-criterion, $x_{i,12,4}$ noise sub-criterion, $x_{i,12,5}$ potential water contamination sub-criterion, and $x_{i,12,6}$ radioactivity sub-criterion. Each sub-criterion will be evaluated with a weight in a range from 1 to 5.

Table 12. Performance values $x_{i,12,j}$ for 12th criterion are defined according to values available in References [31,36–39].

Subsidence v_h *(mm/year)*	$v_h \geq 100$	$100 > v_h \geq 60$	$60 > v_h \geq 40$	$40 > v_h \geq 20$	$v_h < 20$
$x_{i,12,1}$	1	2	3	4	5
Potential Seismicity *PGV (cm/s)* *PGA (cm/s²)*	$0.5 \leq$ PGV ≤ 1.6 $9 \leq$ PGA ≤ 43	$0.2 \leq$ PGV ≤ 0.6 $4 \leq$ PGA ≤ 18	$0.07 \leq$ PGV ≤ 0.23 $1.5 \leq$ PGA ≤ 7.3	$0.03 \leq$ PGV ≤ 0.09 $0.6 \leq$ PGA ≤ 3	$0.01 \leq$ PGV ≤ 0.02 $0.2 \leq$ PGA ≤ 1.2
$x_{i,12,2}$	1	2	3	4	5
Land Use *(m²/kW)*	LUI > 40	$40 \geq$ LUI > 30	$30 \geq$ LUI > 20	$20 \geq$ LUI > 10	LUI ≤ 10
$x_{i,12,3}$	1	2	3	4	5
Noise *(dB)*	dB ≥ 100	$100 >$ dB ≥ 90	$90 >$ dB ≥ 80	$80 >$ dB ≥ 70	dB < 70
$x_{i,12,4}$	1	2	3	4	5
Potential Water Contamination *TDS (mg/L); (pH)*	TDS ≥ 1200; pH ≤ 3 or pH > 8.5	$900 \leq$ TDS < 1200; $3 <$ pH < 4	$600 \leq$ TDS < 900; $4 \leq$ pH < 5	$300 \leq$ TDS < 600; $5 \leq$ pH < 6.5 or $7.5 <$ pH ≤ 8.5	TDS < 300; $6.5 \leq$ pH ≤ 7.5
$x_{i,12,5}$	1	2	3	4	5
Radioactivity *(ppm)*	$2 \leq K^{40} < 6$ $1 \leq$ Th ≤ 25 $1 \leq$ U ≤ 7	$8 \leq$ Th ≤ 18 $1.5 \leq$ U ≤ 5.5	$0.2 \leq K^{40} < 2$ $0.5 \leq$ Th ≤ 10 $0.2 \leq$ U ≤ 0.4	$0.7 \leq K^{40} < 3.8$ $0.7 \leq$ Th ≤ 3.8 $0.2 \leq$ U ≤ 0.6	$0 \leq K^{40} < 2$ $0.1 \leq$ Th ≤ 7 $0.1 \leq$ U ≤ 9
$x_{i,12,6}$	1	2	3	4	5
Total Environmental Impact	AV ≤ 1	$1 <$ AV ≤ 2	$2 <$ AV ≤ 3	$3 <$ AV ≤ 4	$4 <$ AV ≤ 5
$x_{i,12}$	1	2	3	4	5

4. Results

In this section, the purpose of the MCDM matrix is discussed and decision-making results are presented. The matrix is used to rank four different geothermal projects for two given end uses mentioned in Section 2.4, only electricity production and direct heat use, respectively. For a successful comparison, the geothermal projects being compared should be targeting the same end use, so that the comparison considers how key technical and economic factors are being utilised. Therefore, the two main scenarios are considered separately for all the projects (four sites in Table 13). In order to be able to execute an overall comparison, all the criterions mentioned in Section 3 are assessed. As stated before, relative importance in the decision making of each criterion is considered equal, meaning each criterion is equally influential on the final decision and therefore w_j, $\forall j$ is set as 1. Final evaluation of chosen sites that includes all criteria is calculated as the average mean of 12 criterions Moreover, the extraction technology examined in this analysis is doublet producer-injector technology, described in Section 2.3. For the four geothermal sites, the input parameters used for Equations (1)–(12) are presented in Table 13.

These input parameters represent realistic geothermal sites. Some of the observed sites are currently in usage for electricity production, aquaculture, agriculture, and food industry. The observed sites are not in the same geological and environmental area so the MCDM tool could be properly used, because it evaluates and compares different geothermal sites at various geological settings for two scenarios.

Table 13. Input parameters for the multiple-criteria decision-making (MCDM) matrix for four selected geothermal sites.

Parameter	Unit	*Site 1*	*Site 2*	*Site 3*	*Site 4*
Brine flow rate, q	L/s	83.33	20	5.5	77.1
Inlet temperature, T_H	°C	170	80	140	80
Outlet temperature, T_C	°C	70	32	70	40
Geothermal gradient	°C/100 m	6.18	6.1	3	6
Number of wells	No.	1	1	1	1
Specific heat capacity, c_p	J/kgK	4185.5	4185.5	4185.5	4185.5
Corrosion and scaling	LSI	1.5	0.5	0.5	0.5
Fluid	kg/m^3	897.3	971.76	925.9	971.76
Th. efficiency, $\eta_{max,ele}$	%	6.18	6.1	3	6
Max. efficiency, $\eta_{max,\ heat}$	%	76.47	62.5	78.57	53.33

[1] Density of the geothermal water changes with the temperature according to Reference [40].

4.1. Economic Parameters

The problems encountered with the development of geothermal energy systems are mostly related to the high upfront costs and the related finances. The high upfront costs are usually caused by the costs involved with the drilling of the wells, as mentioned in Section 2.2. The drawback with financing geothermal projects relates to the substantial uncertainties in the performance of the wells. Namely, EGS technology is still in an R&D phase since only a handful of projects have been realized. With estimated CAPEX and OPEX values, the LCOE of each site is calculated (Table 14), according to Equation (13) and passed on to the MCDM matrix. For analysis in this study, 55% of the CAPEX is excluded, since all the selected geothermal sites do not require drilling phase, i.e., the geothermal injector-producer doublet exist. Moreover, CAPEX is represented with specific investment costs in €/MW and OPEX is calculated as the percentage of the CAPEX in the range between 7–20%, depending on the year of operation.

$$LCOE = \frac{\sum_{t=1}^{T} \frac{CAPEX_t + OPEX_t}{(1+r)^t}}{\sum_{t=1}^{T} \frac{E_t}{(1+r)^t}}, \tag{13}$$

where E_t represents the amount of the electricity or heat produced in the year t. r represents the discount rate. T is the lifetime of the project.

Table 14. Economic parameters used for levelized cost of energy (LCOE) calculations.

Parameter	Unit	Value
Specific costs	€/MW	4,500,000; 6,750,000 [1]; 9,000,000
OPEX	€/MWh	10; 20 [1]; 30
Discount	%	6; 7 [1]; 8.7
Lifetime	years	20; 30 [1]; 40
Energy produced, E_t	MWh (annual)	Calculated [2]

[1] Base case values are used in the base case scenario, left values represent decreased and right values increased values compared to the base case scenario that is used in the sensitivity analysis; [2] The annual produced electricity or heat is calculated as a product of Equation (2) and (3) or Equation (4).

4.2. First Scenario—Electricity Production Only

This scenario includes the end user option for only electricity production. The results of running the modelled MCDM matrix on the four selected geothermal sites to rank their feasibility according to economic-environment-social assessment are illustrated in Table 15. To estimate each criterion, the parameters from Table 13 and from Table 14 the base case values are used.

Table 15. Ranking of the selected geothermal sites in this study for electricity production.

Criterion	Site 1	Site 2	Site 3	Site 4
$x_{i,1}$	4	1	1	1
$x_{i,2}$	1	2	4	1
$x_{i,3}$	3	1	2	1
$x_{i,4}$	5	5	3	5
$x_{i,5}$	5	2	4	2
$x_{i,6}$	1	1	1	1
$x_{i,7}$	1	4	4	4
$x_{i,8}$	4	5	4	5
$x_{i,9}$	5	5	5	5
$x_{i,10}$	3	3	3	3
$x_{i,11}$	5	5	5	5
$x_{i,12}$	5	5	5	5
Final	3.5	2.83	3	2.75

4.3. Second Scenario—Heat Production Only

This scenario includes the end user option for only heat production, for direct use applications. The results of running the modelled MCDM matrix on the four selected geothermal sites for this scenario to rank their feasibility according to economic-environment-social assessment are illustrated in Table 16. To estimate each criterion, the parameters from Table 13 and from Table 14 base case values are used. Load factor for specified sites is defined according to the application of the heat and as follows. It is modelled that all the sites use the obtained heat for district heating and accordingly the load factor is set to 0.3.

Table 16. Ranking of selected geothermal sites in this study for heat production.

Criterion	Site 1	Site 2	Site 3	Site 4
$x_{i,1}$	4	5	4	4
$x_{i,2}$	1	2	3	1
$x_{i,3}$	5	5	5	5
$x_{i,4}$	5	5	3	5
$x_{i,5}$	5	2	4	2
$x_{i,6}$	4	4	4	4
$x_{i,7}$	1	4	4	4
$x_{i,8}$	3	4	3	4
$x_{i,9}$	2	2	2	2
$x_{i,10}$	1	1	1	1
$x_{i,11}$	4	5	4	5
$x_{i,12}$	5	5	5	5
Final	3.33	3.67	3.5	3.5

4.4. Sensitivity Analysis

Sensitivity analysis was carried out to see how the change of independent variables (CAPEX, discount rate, OPEX, lifetime) affects the dependent variable, i.e., system LCOE, under certain conditions. A sensitivity analysis makes it possible to distinguish between high-leverage variables, whose values have a significant impact and low-leverage variables, whose values have minimal impact. Site 1 and scenario of only electricity production were chosen for sensitivity analysis for the system LCOE. Values of observed input parameters used in the sensitivity analysis are shown in Table 17.

Table 17. The sensitivity of the system levelized cost of energy (LCOE) to input parameters.

Input Parameter	Unit	Decrease	Base	Increase
CAPEX	€/MW	4,500,000	6,750,000	9,000,000
Discount rate	%	6	7	8
OPEX	€/MWh	20	30	40
Lifetime	years	20	30	40

Tornado plot showing the sensitivity of calculated LCOE to changes in a selection of parameters is shown in Figure 7. The central value of the plot represents the calculated LCOE for the base case with the amount of 93 €/MWh. It could be seen that the most influence on system LCOE has CAPEX, followed by OPEX, discount rate and the lifetime of the project, respectively. Lowering the CAPEX, which consist mostly of the drilling costs, is crucial for making the EGS geothermal sites commercially available and competitive with other renewable sources.

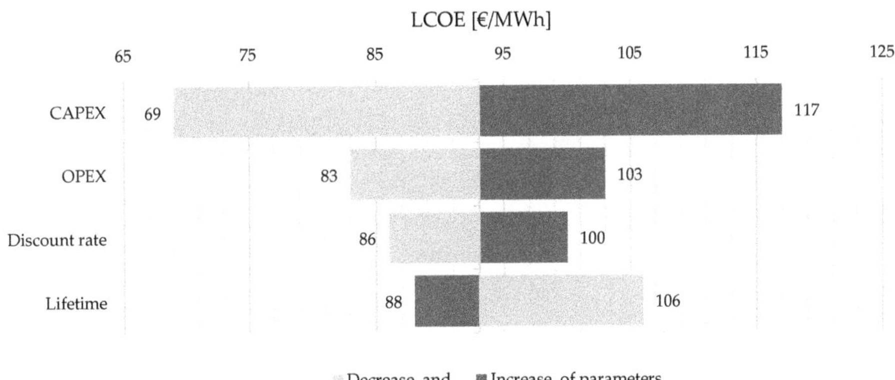

Figure 7. Tornado plot showing the sensitivity of the calculated levelized cost of energy (LCOE) to changes in a selection of parameters.

4.5. Discussion

The presented MCDM as a subprocess of the DMS-TOUGE will enable a comprehensive understanding of the interaction between economic, geological, social, and technical uncertainty. The analysis conducted in this paper is a first order analysis and will benefit from additional work gained from the MEET project. It is expected that the model will receive several modifications in order to make it more accurate and impactful. As it can be seen from the final evaluation of each site in Table 15, the temperature and the flow rate are one of the most important constraints for development of an EGS project, especially in the case of electricity production only. Site 1 presents the highest temperature of the produced geothermal brine, and the highest flow rate. These conditions lead to the greater installed power at this site, compared to the others, and thereby the largest amount of electricity produced. When comparing the results obtained for two given scenarios of end-use, it is clear that the electricity generation is a better solution for the Site 1, since the final evaluation for this scenario is 3.5, compared to the scenario of heat production with a final evaluation of 3.33. The selected site is built to produce electricity instead of heat. It can be concluded that in most cases, higher reservoir temperature and hence higher temperature at the wellhead are suitable and economically viable for electricity production. Sometimes, specific end-user requirements, for example, heat load for industry regardless of the very expensive pipeline system, may determine final end user application. This can be manipulated by DM via weight factors settings. Moreover, for Site 2 in the scenario of electricity production, the evaluation is 2.83, that is, significantly lower than for the scenario of heat production with evaluation of 3.67. The decision for this site between the two scenarios is in favour of direct use heat production since, as it can be seen from the parameters from Table 13, the inlet temperature, as well as the difference between inlet and outlet temperature, is smaller than that for Site 1 or Site 3. In addition to this, the fluid flow rate for Site 2 is notably lower compared to Site 1 and Site 3. Therefore, the results obtained with the proposed MCDM matrix demonstrate that using low-to-moderate-temperature geothermal resources in the direct-heat applications, given the right conditions, is an economically feasible business as stated in Reference [41]. As stated before, practical results can be altered according to end user needs. For Site 3, the final evaluation for the electricity generation is 3 and 3.5 for the heat production scenario. Although the inlet temperature is quite high, the lowest fluid flow rate of all selected sites affects the final evaluation, especially in the scenario of electricity production. Site 4 has the lowest evaluation of 2.75 in the scenario of electricity generation. Namely, this site has the closest difference between the temperature of the produced geothermal fluid and the injection temperature, leaving thereby a small amount of heat that can be transferred to the secondary working fluid. The smaller the difference between those temperatures, the smaller the amount of produced electricity. However, in the case of the second scenario of heat production, it is visible that this site has a better final performance indicator. Therefore, lower reservoir temperature and a smaller difference between wellhead inlet and outlet temperature are more suitable for direct heat use applications. Social and environmental impact of selected EGS sites are favorable for both scenarios and therefore have high-performance indicators for those criteria. Namely, the installed power of the selected sites imposes modest land-use intensity, considering that technical characteristics of the EGS systems atmospheric emissions are zero. Moreover, selected air-cooled systems do not require additional cooling water, leading to no surface water exploitation. The social impact is positive, since the selected projects are relatively small to medium size projects, and represent more benefits than disadvantages at the local level since the balance between power plant size and local impact and benefits in terms of FT and C&M jobs is established.

5. Conclusions

This paper presents the concept of multi-scale Decision-Making Support Tool for Optimal Usage of Geothermal Energy (DMS-TOUGE) and one of its main features, the multiple-criteria decision-making. These features are used for economic and environmental assessment of enhanced geothermal systems projects. The paper contributes with expanded and detailed criteria related to environmental and social impact giving the necessary emphasis on so far neglected important aspects for successful completion of geothermal projects. Moreover, by combining technical, economic, environmental, and social aspects of geothermal projects, the multiple-criteria decision-making presented in this work gives a comprehensive assessment of enhanced geothermal systems projects. The multiple-criteria decision-making (MCDM) matrix is tested on the case of four operating demonstration geothermal sites where it is used for assessment and comparative analysis. In the application of MCDM matrix, the choice of the weight to assign to the different criterions affects the evaluation of the project. In this work, it is considered that each criterion has equal relative importance in decision making, therefore the weights, w_j, $\forall j$, associated with each criterion, are considered identical and are valued with unitary weight. It is, however, important to highlight that the weights assigned to each criterion can vary depending on the DM's standpoint. Therefore, the results of the final evaluation of the project will vary if a greater weight is assigned to specific criterions according to DM's assessment. Further steps in tool development are the incorporation of many other relevant decision-making input parameters, not included in the analysis made in this study, testing its capabilities and subsequent verification and validation of the tool. Its proper function is of key importance for work packages and several deliverables of H2020 project such as generation of several layers for mapping (of the main promising European sites where EGS can or should be implemented in a near future) in different resolutions: EU wide layer, layer considering different geologic features, and pilot site layer. The multi-scale

Decision-Making Support Tool for Optimal Usage of Geothermal Energy (DMS-TOUGE) will be fully developed as part of the Horizon 2020 project: Multidisciplinary and multi-context demonstration of EGS exploration and Exploitation Techniques and potentials (MEET, GA No 792037). The developed tool will be useful for the decision makers involved in enhanced geothermal system projects associated with applications to nearly unexploited reservoir types (Variscan orogenic belt). This will provide valuable information to decision makers and investors for assessment of enhanced geothermal systems projects considering site-specific environmental, techno-economic and geological features.

Author Contributions: S.R., P.I. and I.R. have contributed to developing the concept, formulation and validation of the methodology. S.R. and P.I. have been involved in the implementation of the problem, and analysis of the results. I.R., T.B. and G.T. have been involved in investigation, supervision and project administration. All the authors are involved in preparing the manuscript.

Funding: This project has received funding from the European Union's Horizon 2020 research and innovation programme under grant agreement No 792037 and support from Department of Energy and Power Systems of University of Zagreb Faculty of Electrical Engineering and Computing.

Conflicts of Interest: The authors declare no conflict of interest.

References

1. IRENA (International Renewable Energy Agency). Renewable Capacity Statistics 2018. Available online: https://www.irena.org/publications/2018/Mar/Renewable-Capacity-Statistics-2018 (accessed on 20 December 2018).
2. Beckers, K.F.; Lukawski, M.Z.; Anderson, B.J.; Moore, M.C.; Tester, J.W. Levelized costs of electricity and direct-use heat from Enhanced Geothermal Systems. *J. Renew. Sustain. Energy* **2014**, *6*. [CrossRef]
3. Alimonti, C.; Falcone, G.; Liu, X. Potential for Harnessing the Heat from a Mature High-Pressure-High-Temperature Oil Field in Italy. In Proceedings of the SPE Annual Technical Conference and Exhibition, Amsterdam, The Netherlands, 27–29 October 2014; pp. 3763–3775. [CrossRef]
4. Zhang, L.; Yuan, J.; Liang, H.; Li, K. Energy from Abandoned Oil and Gas Reservoirs. In Proceedings of the SPE Asia Pacific oil and gas Conference and Exhibition, Perth, Australia, 20–22 October 2008; pp. 404–503.
5. Davis, A.P.; Efstathios, E.; Michaelides, E. Geothermal power production from abandoned oil wells. *Energy* **2009**, *34*, 866–872. [CrossRef]
6. Bu, X.; Ma, W.; Li, H. Geothermal energy production utilizing abandoned oil and gas wells. *Renew. Energy* **2012**, *41*, 80–85. [CrossRef]
7. Barbacki, A.P. The use of abandoned oil and gas wells in Poland for recovering geothermal heat. In Proceedings of the World Geothermal Congress, Kyzshu-Tohoku, Japan, 28 May–10 June 2000.
8. Xin, S.; Liang, H.; Hu, B.; Li, K. Electrical Power Generation from Low Temperature Co-Produced Geothermal Resources at Huabei Oilfield. Proceedings Thirty-Seventh Workshop on Geothermal Reservoir Engineering, Stanford, USA, 2012. Available online: https://pangea.stanford.edu/ERE/pdf/IGAstandard/SGW/2012/Xin.pdf (accessed on 26 April 2019).
9. Cheng, W.; Li, T.; Nian, Y.; Wang, C. Studies on geothermal power generation using abandoned oil wells. *Energy* **2013**, *59*, 248–254. [CrossRef]
10. Zhang, L.; Liu, M.; Kewen, L. Estimation of Geothermal Reserves in Oil and Gas Reservoirs. In Proceedings of the SPE Western Reginal Meeting, San Jose, CA, USA, 24–26 March 2009.
11. Guercio, M.; Bonafin, J. The Velika Ciglena Geothermal Binary Power Plant. In Proceedings of the 6th African Rift Geothermal Conference, Addis Ababa, Ethiopia, 2–4 November 2016.
12. Tipurić, D.; Bruketa, N.; Ćenan, D. *Izvodljivost Programa Gospodarske uporabe Geotermalne Energije na Lokaciji Velika Ciglena*; Sveučilište u Zagrebu, Ekonomski fakultet Zagreb: Zagreb, Croatia, 2007.
13. Golub, M.; Kurevija, T.; Koščak-Kolin, S. Thermodynamic cycle optimization in the geothermal energy production. *Rud-Geol.-Naft. Zagreb* **2004**, *16*, 81–84.
14. Lu, S.M. A global review of enhanced geothermal systems (EGS). *Renew. Sustain. Energy Rev.* **2017**, *81*, 2902–2921. [CrossRef]
15. Sanyal, S.K.; Morrow, J.W.; Butler, S.J.; Robertson-Tait, A. Is EGS Commercially Feasible? *GRC Trans.* **2007**, *31*. Available online: https://catalog.data.gov/dataset/is-egs-commercially-feasible/resource/6dde9e1b-8cf1-4b4c-96ab-28a8ed25eba9 (accessed on 26 April 2019).

16. Sanyal, S.K.; Morrow, J.W.; Butler, S.J.; Robertson-Tait, A. Cost of electricity from enhanced geothermal systems. In Proceedings of the Thirty-Second Workshop on Geothermal Reservoir Engineering, Stanford, CA, USA, 22–24 January 2017.
17. Clark, C.; Sullivan, J.; Harto, C.; Han, J.; Wang, M. Life cycle environmental impacts of geothermal systems. *J. Renew. Sustain. Energy* **2013**, 5. [CrossRef]
18. Bayer, P.; Rybach, L.; Blum, P.; Brauchler, R. Review of life cycle environmental effects of geothermal power generation. *Renew. Sustain. Energy Rev.* **2013**, *26*, 446–463. [CrossRef]
19. Olasolo, P.; Juárez, M.C.; Olasolo, J.; Morales, M.P.; Valdani, D. Economic analysis of Enhanced Geothermal Systems (EGS). A review of software packages for estimating and simulating costs. *Appl. Therm. Eng.* **2016**, *104*, 647–658. [CrossRef]
20. Office of Energy Efficiency and Renewable Energy, Geothermal Electricity Technology Evaluation Model (GETEM). Available online: https://www.energy.gov/eere/geothermal/geothermal-electricity-technology-evaluation-model (accessed on 15 October 2018).
21. Heidinger, P.; Dornstadter, A.; Fabritius, A. HDR economic modelling HDRec software. *Geothermics* **2006**, *28*, 71–683. [CrossRef]
22. De Jesus, A.C. Environmental benefits and challenges associated with geothermal power generation. In *Geothermal Power Generation: Developments and Innovation*; R. DiPippo; Woodhead Publishing, 2016; pp. 477–498. [CrossRef]
23. Cataldi, R. Social Acceptability of Geothermal Energy: Problems and Costs. 1997, pp. 343–351, Ankara. Available online: https://www.geothermal-energy.org/pdf/IGAstandard/ISS/2001Romania/cataldi.pdf (accessed on 26 April 2019).
24. Popovski, K. Political and Public Acceptance of Geothermal Energy. IGC2003. Reykjavík, Iceland, September 2003. Available online: https://orkustofnun.is/gogn/unu-gtp-report/UNU-GTP-2003-01-03.pdf (accessed on 26 April 2019).
25. Shang, Y. Resilient Multiscale Coordination Control against Adversarial Nodes. *Energies* **2018**, *11*, 1844. [CrossRef]
26. Rockafellar, R.T.; Uryasev, S. Conditional value-at-risk for general loss distribution. *J. Bank. Financ.* **2002**, *26*, 1443–1471. [CrossRef]
27. Ilak, P.; Rajšl, I.; Đaković, J.; Delimar, M. Duality Based Risk Mitigation Method for Construction of Joint Hydro-Wind Coordination Short-Run Marginal Cost Curves. *Energies* **2018**, *11*, 1254. [CrossRef]
28. Ilak, P.; Krajcar, S.; Rajšl, I.; Delimar, M. Pricing Energy and Ancillary Services in a Day-Ahead Market for a Price-Taker Hydro Generating Company Using a Risk-Constrained Approach. *Energies* **2014**, *7*, 2317–2342. [CrossRef]
29. Gehringer, M.; Loksha, V. *Geothermal Handbook: Planning and Financing Power Generation (English)*. ESMAP technical report; no. 002/12. 2012. Available online: https://www.esmap.org/sites/esmap.org/files/DocumentLibrary/FINAL_Geothermal%20Handbook_TR002-12_Reduced.pdf (accessed on 7 January 2019).
30. Mengying, L.; Noam, L. Comparative Analysis of Power Plant Options for Enhanced Geothermal Systems (EGS). *Energies* **2014**, *7*, 8427–8445. [CrossRef]
31. Soldo, E.; Alimonti, C. From an oilfield to a geothermal one: use of a selection matrix to choose between two extraction technologies. In Proceedings of the World Geothermal Congress 2015, Melbourne, Australia, 19–25 April 2015.
32. Moon, H.; Zarrouk, S.J. Efficiency of geothermal power plants: A worldwide review. In Proceedings of the New Zealand Geothermal Workshop 2012, Auckland, New Zealand, 19–21 November 2012.
33. AL-Mahrouqi, J.; Falcone, G. An Expanded Matrix to Scope the Technical and Economic Feasibility of Waste Heat Recovery from Mature Hydrocarbon Fields. In Proceedings of the 41st Workshop on Geothermal Reservoir Engineering, Stanford, CA, USA, 22–24 February 2016.
34. Rafferty, K. Industrial Processes and the Potential for Geothermal Applications. Report. GHC Bulletin: Oregon, 2003. Available online: https://oregontechsfstatic.azureedge.net/sitefinity-production/docs/default-source/geoheat-center-documents/quarterly-bulletin/vol-24/art21683ee4362a663989f6fff0000ea57bb.pdf?sfvrsn=8b258d60_4 (accessed on 26 April 2019).
35. Cataldi, R. Social Acceptability of Geothermal Energy: Problems and Costs. EC International Geothermal Course. Oradea, Romania, 2001; pp. 343–351. Available online: https://www.geothermal-energy.org/pdf/IGAstandard/ISS/2001Romania/cataldi.pdf (accessed on 26 April 2019).

36. Zang, A.; Oye, V.; Jousset, P.; Deichmann, N.; Gritto, R.; McGarr, A.; Majer, E.; Bruhn, D. Analysis of induced seismicity in geothermal reservoirs—An overview. *Geothermics* **2014**, *52*, 6–21. [CrossRef]
37. Johansson, T.B.; Patwardhan, A.; Nakicenovic, N. Gomez-Echeverri, Land-use and Energy Systems. In *L. Global Energy Assessment: Toward a Sustainable Future*; Cambridge University Press: Cambridge, UK, 2012; pp. 232–234.
38. DiPippo, R. Geothermal energy Electricity generation and environmental impact. *Energy Policy* **1997**, *19*, 798–807. [CrossRef]
39. Johnson, S. Natural radiation. *Va. Miner.* **1991**, *37*, 9–16.
40. Zongjun, G.; Yonggui, L.; Yaun, G. The Principle of Density Differences Drive Geothermal Water to Move and the Short-Range Recharge Model of Geothermal Water in Hilly Area. In Proceedings of the Thirty-Ninth Workshop on Geothermal Reservoir Engineering PROCEEDINGS, Stanford, CA, USA, 24–26 February 2014.
41. Lund, J.W. Direct Utilization of Geothermal Energy. *Energies* **2010**, *3*, 1443–1471. [CrossRef]

Article

Importance of Reliability Criterion in Power System Expansion Planning

Goran Slipac [1,*], Mladen Zeljko [2] and Damir Šljivac [3]

1 Croatian Power Company (HEP d.d.), Zagreb 10000, Croatia
2 Energy Institute Hrvoje Požar (EIHP), Zagreb 10000, Croatia; mzeljko@eihp.hr
3 Faculty of Electrical Engineering, Computer Science and Information Technology (FERIT), Josip Juraj Strossmayer University of Osijek, Osijek 31000, Croatia; damir.sljivac@ferit.hr
* Correspondence: goran.slipac@hep.hr

Received: 15 March 2019; Accepted: 30 April 2019; Published: 7 May 2019

Abstract: The self-sufficiency of a power system is no longer a relevant issue at the electricity market, since day-to-day optimization and security of supply are realized at the regional or the internal electricity market. Research connected to security of supply, i.e., having reliable power capacities to meet demand, has been conducted by transmission system operators. Some of the common parameters of security of supply are loss of load probability (LOLP) and/or loss of load expectation (LOLE), which are calculated by a special algorithm. These parameters are specific for each power system. This work presents the way of calculating LOLP as well as the optimization algorithm of LOLP, which takes into consideration the particularities of the power system. It also presents a difference in the treatment of LOLP regarding the observed power system and the necessary installed power capacity if applied to the calculated LOLP in relation to the optimized LOLP. As a conclusion, the study analyzed the parameters impact the regional electricity market—where the participants are countries with different development levels and various particularities of power systems—i.e., what it means when the same LOLP criterion is applied to them and the optimized LOLP.

Keywords: security of supply; LOLP optimization; VOLL; generation expansion planning; reliability

1. Introduction

The electricity market and the security of electricity supply are defined by several directives and regulations. One of the most important documents stipulating measures for secure customer security is the 2005/89/EC Directive [1], whereby the energy union concept is based on a few (eight in total) new bills, a new regulation on the preparation for risks being among them, which should replace said 2005/89/EC Directive. The analysis of the adequacy of power systems is conducted by the transmission system operators on the basis of methodological instructions by ENTSO-e, and that report [2] serves as the basis for the assessment of mid-term security of supply of the power system. The basic probability indicators for drawing up the mid-term adequacy of power systems are loss of load probability (LOLP) and loss of load expectation (LOLE). LOLP in a power system is expressed as a percentage of time, annually or monthly, in which the power system will not be able to meet demand with available generation capacities. LOLE is the index on which the basis of LOLP shows the total number of hours in a year, in which the power system will not be able to meet demand. In this paper, reliability calculations are described, followed by examples of generation expansion of a real power system under different reliability indices. As it is stated in [3], it is necessary to quantify the value of lost load (VOLL) to obtain the economic value of adequacy. In order to put a value on reliability, a commonly employed parameter is the VOLL, which measures the damage suffered by consumers when the supply is curtailed. In cases of productive activities (such as industrial processes engaged in the production of a good), an objective measure of the cost of interruptions based on the loss of production or the linked benefit is possible.

At the moment in the European Union (EU), a new Regulation (EU) 2018/2019 of the European parliament and of the Council of risk-preparedness in the electricity sector is being prepared, and repealing Directive 2005/89/EC is supposed to be adopted at the beginning of 2020. As it is described in the new Regulation, the electricity sector in the EU is undergoing a profound transformation, characterized by more decentralized markets with more players, a higher penetration of renewable energy, and better interconnected systems. The main idea of the internal electricity market is that well-functioning markets and systems with adequate electricity interconnections are the best guarantee of security of supply.

A common approach to electricity crisis prevention and management also requires that member states use the same methods and definitions to identify risks relating to the security of supply and are in a position to effectively compare how well they and their neighbors perform in that area. This Regulation identifies two indicators for monitoring the security of electricity supply in the Union—expected energy non-served (EUE), expressed in GWh/year, and LOLE, expressed in hours per year. Those indicators are part of the European resource adequacy assessment carried out by ENTSO-E. In the Regulation, there is a description of the methodology of calculation of security of supply. To ensure the coherence of risk assessments in a manner that builds trust between member states in an electricity crisis, a common approach to identifying risk scenarios is needed. Therefore, after consulting the relevant stakeholders, ENTSO-E should develop and update a common methodology for risk identification in cooperation with the Agency for the Cooperation of Energy Regulators (ACER) and with the Electricity Coordination Group (ECG), which was set up by a Commission decision on 15 November 2012 as a forum in which to exchange information and foster cooperation among member states—in particular, in the area of security of supply—in its formation composed only of representatives of the member states. ENTSO-E should propose the methodology, and ACER should approve it. When consulting the ECG, ACER is to take utmost account of its views. ENTSO-E should update the common methodology for risk identification where significant new information becomes available.

ENTSO-e conducts studies about Scenario Outlook and Adequacy Forecast, trying to calculate a Pan-European power system adequacy. In their report [2], there is a question raised related to capacity adequacy. If the reliability index is met beyond the national legal requirement, how large is the related capacity margin surplus available for the neighboring area, or even system wide? Additionally, if the reliability index is not met, by how much would the available capacity need to increase to meet the target? Another question is about the methodology applied.

The LOLP index is used in the analyses of generation parts of power systems but also in reliability analyses of both the transmission and the distribution system. There are many more papers regarding the use of VOLL for both transmission and distribution system analysis. VOLL is an essential parameter for transmission system reliability management [4]. Theoretical analysis shows that using more detailed VOLL data allows for better-informed transmission reliability decisions. Incorporating variations and uncertainties into expansion planning produces a plan that has better economy and reliability performance during operations. Recent progress on robust optimization techniques can consider uncertain parameters in expansion optimization [5].

There are also some papers and studies regarding planning a reserve margin, which is a common metric used in generation planning to determine an electric utility's resource need above the typical annual peak load. As a proxy for system reliability, the planning reserve margin is useful in informing resource decisions between detailed reliability studies [6]. Questions about providing electricity with adequate power quality and reliability at a reasonable rate are becoming more important. Therefore, there are a number of country-specific analyses, such as German security of supply or questions about capacity market [7,8]. Of high interest in the field of reliability of power systems or the security of supply is a capacity market or remuneration mechanism. In Reference [9], the analysis begins with identifying the optimum level of investments to the 2030 horizon to ensure capacity adequacy, assuming a reference scenario of EU energy system developments, followed by a simulation of the operation of a wholesale energy-only market to estimate the ability of these investments in order to recover their capital cost. The amount of capacity remuneration required to cover for the missing revenue margin is estimated, as are the implications from hypothetical asymmetric remuneration across the member-states of the EU. Also in Reference [10], there is a question raised regarding the competitive energy market and whether it will be sufficiently addressed in the capacity market to garner adequate resources to keep the lights on. The European electricity market has been constructed as an "energy only market" without the compensation for the costs of generation adequacy. Such a market is able to operate in the presence of the capacity excess. However, gradually, the power reserve margins have been diminishing, causing a fear of power balancing in the coming years [11]. All those papers are dealing with one common question—security of supply or capacity adequacy? The basic methodology always relies on a probabilistic approach.

The consequence of capacity shortage is the reduction in electricity supply, which results in a certain VOLL or energy for customers. The reserve of a power system in all three technological parts—generation, transmission, and distribution—depend on the level of damage, i.e., the specific level of value of undelivered kWh of electricity. Both LOLP and LOLE indices are calculated via an algorithm in which the equivalent curve of load duration is formed by the method of convolution [3–6] from the load duration curve, taking into consideration the probability of outage of every individual generation facility as well as the outage cost, which is VOLL times unserved energy. The fundamental thesis in this paper states that LOLP, and thus LOLE, should not be calculated, but may be optimized regarding VOLL and the available technologies for new generation facilities. The paper further analyzes power systems with different optimum levels of LOLP, which participate in the electricity market. The main objective of this paper is to better understand reliability by the optimization of LOLP and LOLE regarding VOLL and available technologies for new generation facilities.

2. LOLP Optimization Methodology Description

Computer models for the analysis and the planning of the power system were developed on algorithms that described certain aspects of power systems; some of them were simulations, some were optimization types, and some models were a combination of the two [12–14]. One of the most famous models used to calculate the LOLP through long-term planning of a power system is the Wien automatic system planning package (WASP), which is used in this analysis as well. The basis of such analysis is load duration curve (LDC) of the power system. When taking into consideration the probability of the first unit outage, the result is the equivalent load duration curve ($ELDC_1$). The result of the outage of two first units is $ELDC_2$, and $ELDC_n$ is the result when the probability of the outage of the last unit is taken into consideration [13,15]:

$$ELDC_n(x) = p_n \cdot ELDC_{n-1}(x) + q_n \cdot ELDC_{n-1}(x - MW_n), \tag{1}$$

where $ELDC_n$ is the equivalent load duration curve after unit n is on the grid; p is the probability that unit n is available; q is the probability that unit n is unavailable; MW_n is the capacity of unit n and x is load. The expected generation for each unit is then:

$$E_n = p_n \cdot T \cdot \int_{a_n}^{b_n} ELDC_{n-1}(x)dx, \tag{2}$$

where E_n is the generation of unit n; T is the time period represented by the load duration curve; $ELDC_{n-1}$ is the equivalent duration curve, which incorporates the effect of the force outages of unit $n-1$; a_n is the system capacity for units 1, 2, 3, ... , $n-1$; and b_n is the system capacity for units 1, 2, 3, ... , n.

For certain installed capacities of all power plants in the power system (ICP), LOLP is the index determining the probability of the occurrence that the load is higher than the sum of the available capacities of all considered power plants. Energy not served (ENS) is electricity that will not be delivered and represents a part below the ELDCN curve limited by ICP on the left side, as it is shown in Figure 1.

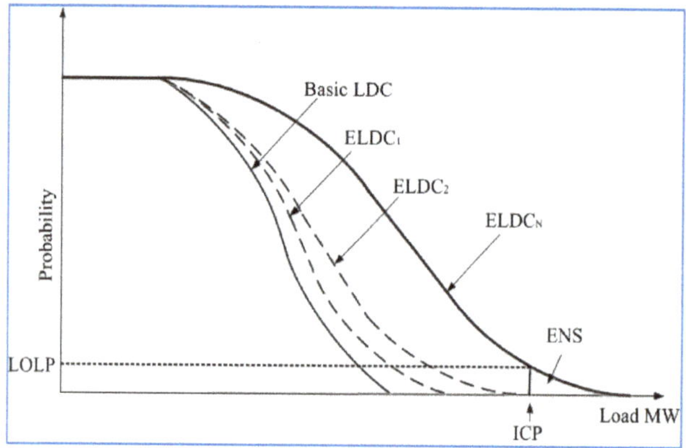

Figure 1. Calculation of the equivalent load duration curve.

To complete the calculation of LOLP, it is necessary to calculate the total cost of the power system, as described in [13,14]. The optimization of the total cost of the power system requires the minimum of a goal function in a specific scenario of the development of the power system. The objective function represents the total cost of the power system, which consists of investment costs for new power plants, fuel costs, operation and maintenance costs (O & M), costs of VOLL, and the remaining value of new power plants after the last planning period year. The LOLP and the LOLE indices are calculated for each year.

$$B_j = \sum_{t=1}^{t=T} \left[I_{j,t} - S_{j,t} + F_{j,t} + M_{j,t} + VOLL_{j,t} \right], \tag{3}$$

B represents the total cost of the power system of the observed scenario; *I* is the capital investment cost in new generation units; *S* is the salvage value of investment costs after the last year of the observed period; *F* represents the fuel costs; *M* represents the non-fuel operation and the maintenance costs; *VOLL* is the outage costs; *t* represents the years of the scenario; *T* is the length of the study period, i.e., the number of years in the scenario; and index *j* is the observed scenario of the generation expansion.

Three main LOLP optimization steps are described as follows:

1. In the first step, one should calculate the power capacity to meet LOLP criteria from 5% to 0.001% on a technical level without any economic consideration. Those values are taken from expert analysis as very high and low boundaries of the LOLP value. The results of calculation for each value of LOLP (starting from five and ending with 0.001%) are the number of MW of installed capacity that should be built in order to meet the criteria of LOLP. Those calculations are conducted under no economical limitations. Thus, the first step is a calculation of how many MW's have to be installed to meet the reliability criteria of LOLP values. That is why LOLP values are first calculated for various combinations of operating power plants. The key point in this stage is to consider reliability aspects only, which means that economic variables are excluded. LOLP criterion is met by introducing new power plants only when it is necessary to satisfy the reliability

criteria regardless of the costs of the power system. The solution is acceptable if the calculated LOLP value is lower or equal to the proposed one for all years of the planned period, i.e.:

$$LOLP \leq LOLP_{UL}, \quad (4)$$

where $LOLP_{UL}$ is the proposed value of LOLP for which the new construction of power plants in the power system is calculated. One can analyze the impact of different technologies or different characteristics of the same technologies of new generation capacity in order to conduct sensitivity analysis.

2. The second step considers economic variables for each value of LOLP or the calculated combination of the necessary power plant capacity for a specific value of LOLP from 5% to 0.001%. In this second step, the objective function B (total cost of the power system) is a function for comparison of different scenarios. It takes into consideration the investment of new capacity, O & M cost, fuel cost, outage cost, salvage value, etc.
3. The third step or case study considers different combinations of power plant capacity for different values of LOLP on a model of a power system. In this paper, it is a Croatian power system, which was taken as a real model. There are four scenarios modeled and compared, two of them with limited LOLP and two of them with no limitation of LOLP.

3. LOLP Optimizing Case Study Analysis

In this paper, two technologies were chosen for power plant candidates—a coal-fired thermal power plant and a natural gas-fired combined gas power plant with usual characteristics taken from literature. Analyses were conducted separately for power plant candidates, coal-fired thermal power plants (TPP), and natural gas-fired combined cycle gas turbine power plants (CCGT). The values of LOLP were simulated for this analysis ranging from 5% to 0.001%. Results are shown in Figure 2.

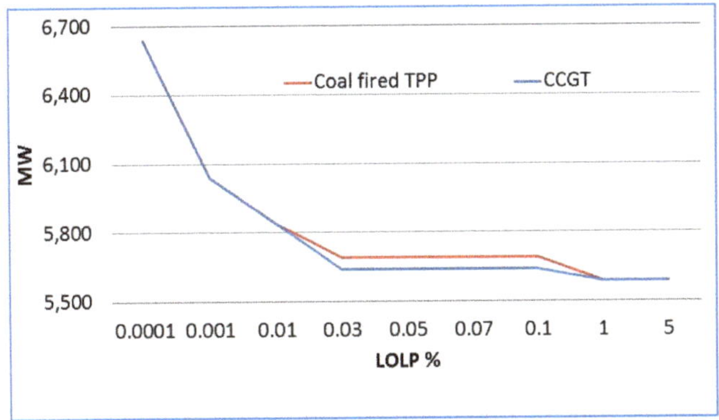

Figure 2. Necessary construction of new capacities to meet loss of load probability (LOLP).

It should be noted that new generation capacity was added in quantities of 20 MW in order to have smooth or continual curves. Characteristics of such small units are the same as regular size units in both analyzed technologies.

After calculating the necessary new constructions of power plants with the given value of LOLP for each LOLP limitation separately, the objective function was calculated for two levels of VOLL—$2 USD/kWh and $10 USD/kWh.

Two levels of VOLL were chosen for analysis, one of $2 USD/kWh and the other of $10 USD/kWh. The former correlated with the gross domestic product (GDP) of around $10,000 USD per capita, and the latter with the GDP of around $40,000 USD per capita [16–18]. The aim was to compare two different levels of economic systems—a developing one and a developed one. It was estimated that there was no need to consider more levels of outage, since this analysis showed differences between higher and lower levels of outage, which corresponded to different developments of economic systems. Other data referring to specific costs of fuel and maintenance were taken from [19,20].

Optimizing the analysis of LOLP started with the calculation of necessary installed power in the observed power system under the limitation of LOLP of 0.0001% (0.5 minutes per year), 0.001% (5.5 min per year), 0.01% (52 min per year), 0.03% (2.5 h per year), 0.05% (4.5 h per year), 0.07% (6 h per year), 0.1% (9 h per year), 1% (3.6 days per year) and 5% (18 days per year) with two levels of outage of $10 USD/kWh and $2 USD/kWh. It should be noted that the same power system was always analyzed, and limitations of LOLP varied just as with VOLL. After obtaining the combination of existing and new power plants, i.e., the total available power for each respective year and for each observed LOLP, the costs of the power system, i.e., the goal function, was calculated as described in the previous chapter. All the values calculated in the goal function refer to the year 2018, which was used as the base in this analysis.

If LOLP was limited at 0.0001%, which is an extreme requirement (usual value amounts at 8 h or 0.1%, e.g., in Ireland), there was a visible increase of necessary capacity or investment costs related to constructing new generation capacities, which was understandable considering the high demand of reliability of the power system. Such a level of security of supply could not be considered necessary due to questionable economic profitability, since the power system disposed a very large reserve (if isolated), thus such a power system would be considered over-capacitated in a techno-economic sense. If such a power system was observed in reality, necessary construction for fulfilling a certain safety level of supply, e.g., of 0.0001%, there would be a need to ensure huge amounts of generation capacities reserve if there were a consideration of not being in the regional electricity market. Where these capacities would be placed in a mathematical sense is not of much importance, except for the fact that they would need to be available, i.e., properly electrically connected.

When the necessary construction of generation capacities of a power system with a coal-fired thermal power plant was analyzed depending on the level of LOLP, we got a curve of necessary construction, as shown in the Figure 3. This curve of necessary construction of generation facilities depending on the LOLP fit well with the theoretical curve. The probabilities (which were extremely small) for which the limitation of LOLP was very small (<0.001), necessary construction of generation capacities grew asymptotically by coming closer to the ordinate axe.

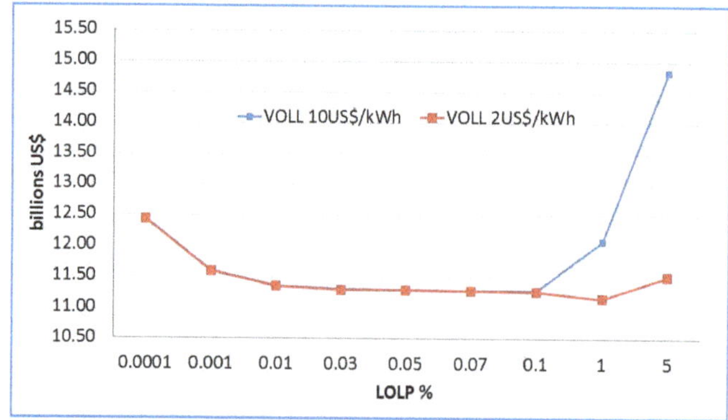

Figure 3. LOLP optimization for two values of lost load (VOLL) levels in the case of a coal-fired thermal power plant (TPP).

If a certain value is added to unsupplied kWh of electricity, then the outage cost becomes dominant in the goal function of the total cost. As a result, the total costs of the power system grow as the level of security of supply decreases. However, more power plant construction leads to a level of higher security of supply. In this case, as more power plant capacities demanded higher investment, the dominant member of the goal function referred to the investment into new power plant capacities, thus the total cost of the power system grew. Therefore, the goal function of the total cost regarding different values of LOLP had the form of a bathtub curve with one more or less expressed as the minimum.

In the cost structure of the power system, it is visible that investment costs, residual value, and fuel costs were the same in both observed cases for VOLL of $2 USD/kWh and $10 USD/kWh starting from LOLP 0.01%, which was understandable since investments for necessary new construction of power plants depended on the given security of supply, and fuel consumption depended on electricity consumption. In the case of VOLL of $10 USD/kWh, the total outage cost as the component of the total cost of the power system increased sharply. It was expected that, in the case of a LOLP value higher than 5%, they would eventually become a dominant part of the total cost of the power system.

These costs were added and presented according to VOLL of $2 USD/kWh and $10 USD/kWh, and the curves of the total costs were obtained. The change of total costs depending on the security of supply in the case of the combined cycle gas turbine (CCGT) only was milder than the case with the coal-fired power plant only. The reasons for this were primarily lower investment costs of new power plants and a lower probability of power plant outages, as shown in Figure 4.

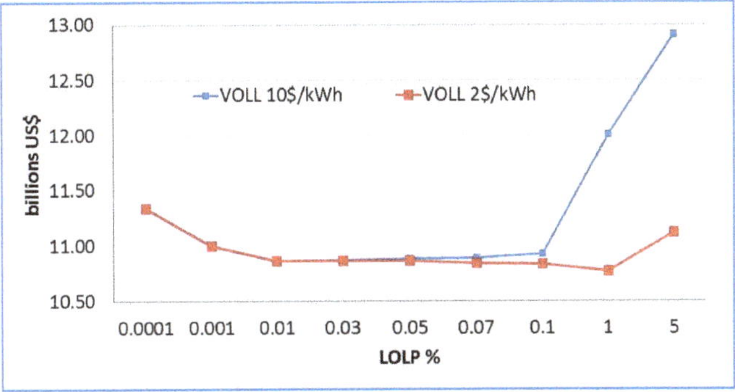

Figure 4. LOLP optimization for two VOLL levels in the case of combined cycle gas turbine (CCGT).

LOLP would accordingly be some kind of necessity or signal for the construction of the power system. In this example, it would particularly refer to the generation capacity, which needed to fit the economic condition of the country, i.e., the power system being analyzed. Optimum LOLP represents not only the criterion for the necessary construction of power plants (i.e., generation capacity) but also an analytic parameter that shows the level of development of the power system and the economic condition as well, i.e., the standard of living of its inhabitants, usually described as GDP per capita. The variable that describes the level of development particularly well is VOLL, and the variable is shown as the one that can connect the level of construction of the power system and the level of economic development. Of course, it is expected that the power system is better constructed, i.e., that more generation facilities are constructed if economic activity is more intensive, i.e., if an economic system can generate a level of additional value that may construct and maintain such a power system to fulfill customers' needs in critical periods. However, if the economic system is not well developed and cannot generate additional value to invest into the energy sector, the power system is then weaker, especially in terms of generation. Consequently, the transmission part of the power system is shown as less reliable and less constructed. On the other hand, electricity and reserve in the power system can be bought at the electricity market. In any case, the construction of generation facilities or the procurement of such facilities, i.e., of electricity at the electricity market, requires additional financial means. An economy that is developed and that has a big additional value is expected to be able to ensure electricity at a considerably higher price in critical periods. The question is whether there will be enough available capacities, which presents a risk. This risk is best described as a dependence on generation facilities, whose work cannot be controlled; without those facilities, there is no safety, long-term or short-term, regarding electricity procurement, amount, dynamics, and quality.

If necessary, power plant construction is a function of LOLP, one gets curves of necessary power, as shown in Figure 5. These curves fit well with theoretical curves of necessary new construction of generation facilities depending on security of supply. In part of the curve with LOLP, from 0.0001% to 0.003%, the difference in necessary construction totals equaled hundreds of MW, as seen in Figure 2. The curve of necessary new construction for the case of the coal-fired TPP showed more value because of less reliability of such facilities in comparison to the gas-fired CCGT. Those differences became negligible for LOLP values higher than one and lower than 0.1 percent.

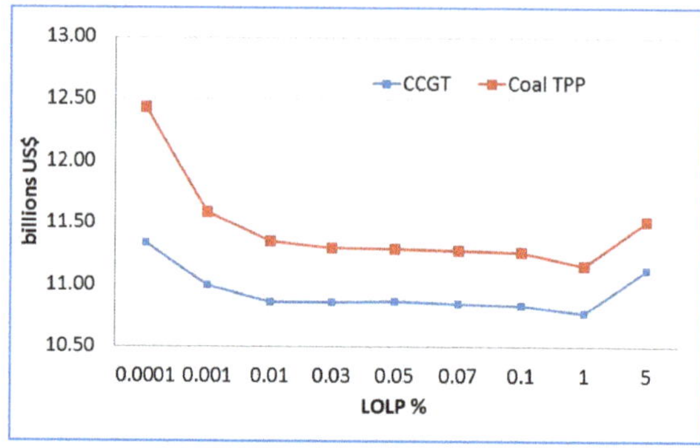

Figure 5. LOLP optimization for VOLL $2 USD/kWh with CCGT and coal fired TPP.

The final (third step) in the determination of the optimal security of supply in comparison to the outage cost was derived by comparing the total costs of the power system for both observed levels of the outage cost. Analysis was carried out with two technologies, the coal-fired TPP and the CCGT.

Figures 5 and 6 show the total costs of the power systems for both observed technologies of the coal-fired TPP and the CCGT depending on the security of supply for the level of VOLL of $2 USD/kWh and $10 USD/kWh. The differences in total costs for the different technologies were expressed at higher levels of customer security of supply, i.e., at lower levels of LOLP when a much higher installed capacity of new power plants was necessary.

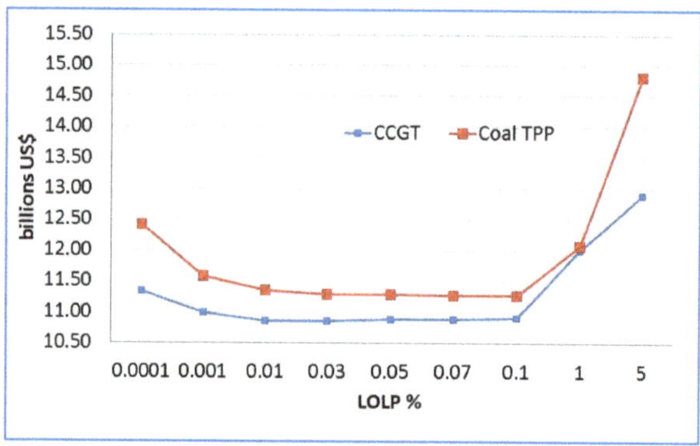

Figure 6. LOLP optimization of LOLP for VOLL $10 USD/kWh with CCGT and coal fired TPP.

Apart from specific characteristics of the analyzed power plant, it is important to see that the VOLL changed the form of the total power system costs curve. For lower levels of VOLL (e.g., for $2 USD/kWh) the diagram of the total cost of the power system is rather symmetrical, while in case of higher VOLL (e.g., for $10 USD/kWh), the curve of total costs clearly expressed minimum.

In conclusion, qualitative analysis of a power system requires the knowledge of economic development levels as well as its perspective. The measure of economic development, which as a variable influences the necessary construction of the power system, is the VOLL. Therefore, we should emphasize that economic development is not the only variable influencing the development of the power system, but it is definitely one of the key variables influencing inhabitants' living standards, traffic intensity, etc. A developed economy enables the development of a power system. Therefore, this is a firm connection between energy, power, and an economic system. The better developed those systems are, i.e., the more synchronized the development of the economic and the power systems is, the narrower the area of optimum construction, i.e., the minimum of total costs, will be.

Interconnected power systems (electrically, regulatory, and economically, which is defined by the electricity market) may count on the power and the energy procured on the market, not only in the case of trading but also in day-to-day operations as a result of operational costs optimization [21–23]. As opposed to self-sufficient power systems in the case of the regional electricity market, the LOLP is not the result but is set as a criterion. Thus, in that case, we may talk about risks managing the possibilities of procuring losses and delivering surpluses. If there are no possibilities of procuring power and energy at the market, and the power system is not developed enough, the LOLP increases and goes into the right part of the diagram, as shown in Figures 5 and 6. In those cases, not only does the procurement of power and energy at the market become important, but the price becomes important as well, which grows due to increased demand in critical periods. Consequently, there is an increase in the total costs of the power systems. Which marginal price of power and electricity is economical or possible to procure power and energy at the market depends on the possibilities of purchase and sale, the economic system development, the available generation capacities, the electricity prices in the regional market, etc. Generally, the more developed and the richer an economic system is, the higher price it can take (Italy, for instance), and vice versa. According to this, the LOLP index expresses the measure of constructiveness of the observed power system, the availability of the generation facilities in the surroundings, and the price of electricity present with the risk of procurement from the market.

Furthermore, in the case of over-construction (installation) of generation facilities in the power system, we arrive to the left part of the diagram, as shown in Figures 5 and 6. These figures display the state of over-construction of the power system if only the observed power system is taken into consideration. Of course, this is just a partially correct view, since power plants (in most cases) are not built to supply one power system but to generate electricity and sell it to customers or to the regional market. The investors build power plants where the final price of electricity is higher, competitive at the market, and where they can sell electricity to customers. However, from the point of view of national energy safety, customer security of supply is higher if the power system is over-constructed, i.e., if there is enough constructed power plants, generation, and other energy facilities, regardless of equivalent working hours. The diagram shows that the minimum of the curve of total costs of the power system moved to the left. The curve's gradient of increase of total costs of the power system for the cases of lower safety levels of customer supply was higher, which means that the increase of total costs due to increase of total damage due to undelivered electricity was more intensive. In practice, this does not mean that power system that has capacity adequacy at its disposal would suffer from reductions. Namely, it is necessary to settle necessary power and energy from neighboring power systems, i.e., at the market of electricity, under the condition that it is available in the amount and the dynamics, such as the price necessary to settle the electricity consumption of the observed power system. In such circumstances, the power system is oriented towards the import of electricity.

The diagrams in Figures 3–6 show for which values of LOLP the minimum observed costs of the power system were achieved. Figure 1 shows the values of LOLP for the observed outage cost and the observed technologies of the new TPP.

4. Case Study

In this analysis, there are four scenarios, as follows:

- SC_0.01% $10—in this scenario, LOLP was limited at 0.01%, VOLL was taken at the level of $10 USD/kWh,
- SC_1% $2—in this scenario, LOLP was limited at 1%, VOLL was taken at the level of $2 USD/kWh,
- SC_LCP $10—in this scenario, LOLP was not limited, the amount of reserve capacity was calculated by the least cost planning (LCP) algorithm, and VOLL was taken at the level $10 USD/kWh,
- SC_LCP $2—in this scenario, LOLP was not limited, the amount of reserve capacity was calculated by the least cost planning algorithm, and VOLL was taken at the level $2 USD/kWh.

All four scenarios presumed the same level and characteristics of electricity consumption and the same characteristics of the new power plants.

Since the field of interest for analysis is defined by values for which the total observed costs are at a minimum, this chapter analyzes the case study scenario of power system development for LOLP values in the field from 1% for the scenario of VOLL of $2 USD/kWh to 0.01% for the scenario of VOLL of $10 USD/kWh. In addition, scenarios of a limited value of LOLP are analyzed. The comparison of these scenarios indicates the essence of the differences between the concepts of the generation expansion planning.

The main results in every scenario presented the number of power plants or the total new capacity becoming operational in the observed period. Although putting power plants into operation was different for each observed scenario, a certain measure of available generation capacity for each scenario was presented by the total construction of the power plants at the end of the observed period necessary to meet certain LOLP requirements. The number of each thermal power plant candidate in a certain year was defined by the number of units that became operational that year.

Load factor of thermal power plants is generally a good indicator of power plant utilization in the power system. Load factors of new power plants, i.e., candidate power plants, were analyzed and aggregated according to technology. Due to the fact that exchange with neighboring power systems was not modeled, the presumption in the analysis was that electricity generation in the thermal and the hydro power plants met the demand. This, of course, meant that the electricity generation was reduced just to meet demand, but it also showed the relation between the necessary generation capacity increases to meet LOLP criteria. Obviously, there was a need for the power plant to work as long as possible to ensure return of investments.

Figure 7 shows the load factor curves of three new thermal power plants, which were analyzed in a scenario with LCP and VOLL of $10 USD/kWh as new thermal power plants, and as such, were put into operation in various years of observed periods.

The left diagram on Figure 7 shows the load factor of the CCGT. The load factor varied regarding other power plants being put in the operation, e.g., the coal-fired TPP. In comparison with the coal-fired TPP, the CCGT had lower investment costs and significantly higher fuel costs, thus the power plant had a lower priority in engagement, and its generation depended on the existing structure of generation in the power system.

As opposed to the 400 MW CCGT, the coal-fired TPP had an even load factor over the years, primarily due to the fact that it had a strong priority in engagement to cover load diagram because of the variable cost of electricity production, e.g., fuel prices. As a case study, the CCGT of 200 MW was set to the same techno-economical characteristic as the one of 400 MW apart from its installed power and technical minimum. It was shown that the load factor for that thermal power plant was even because it was the thermal power plant of lower capacity, thus its relative oscillations of total generation were lower than with a combined gas-fired thermal power plant of 400 MW.

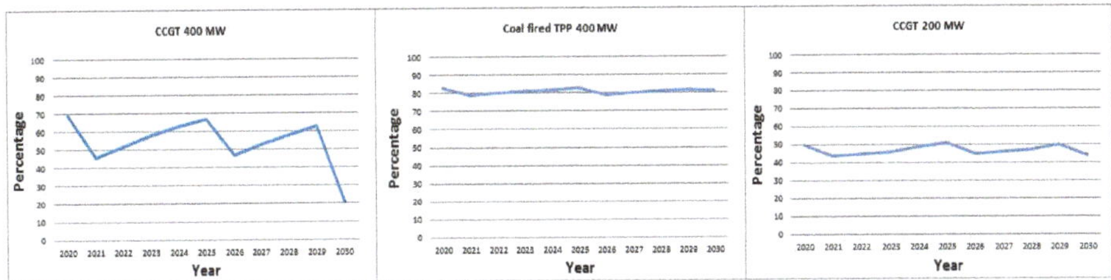

Figure 7. Load factor of new power plants in a scenario with the least cost planning (LCP) algorithm and VOLL of $10 USD/kWh.

This scenario was characterized by a LOLP diagram and a diagram of outage cost per years, as seen in Figure 8. Outage cost (OC) is a kind of damage suffered by consumers (industrial, service, households, etc.) when the supply is curtailed because the power system is not able to meet demand. Outage cost is the total cost, and VOLL is the specific cost, usually expressed in currency over kWh of electricity not delivered to customers. In our study, the first third of the observed period was significant, since the existing state of construction of the generation facilities was adequate to fulfill the security of supply criteria. The maximum value LOLP reached was 0.63% in the years after decommissioning the power plants that were operational in the planned period.

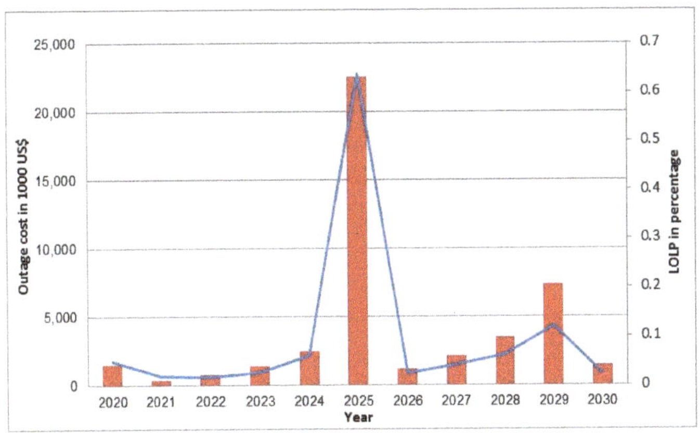

Figure 8. LOLP and outage costs (OC) in a scenario with LCP and VOLL $10 USD/kWh.

In the scenario in which LOLP was given at the level of 0.01%, putting power plants into operation was somehow different, just as the total construction at the end of the observed period was different.

Load factors of new power plants had a decreased gradient towards the end of the observed period. This was understandable if we consider that more power plants were put into operation due to increased levels of customer security of supply along with the similar electricity consumption. This meant that, in this construction scenario, it was a priority to engage power, as seen in Figure 9.

The curve of the value of LOLP and the outage cost per year in the observed planned period showed significant variations, but remained within the framework given by limiting the value to 0.01%. Of course, oscillations depended on putting new power plants into operation and thereby a much higher necessary installed power. All the curves are shown in Figure 10.

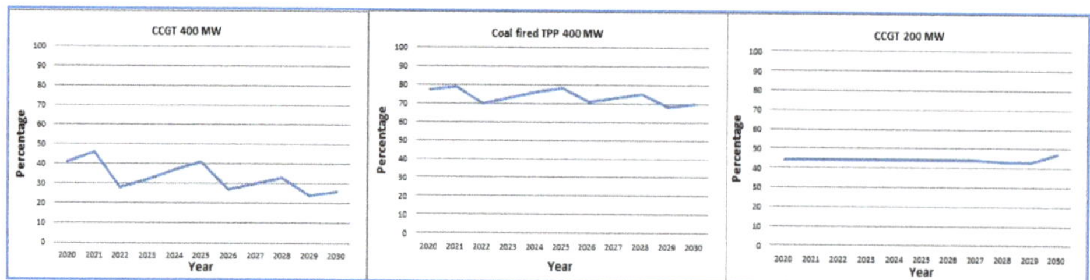

Figure 9. Load factor of new power plants in scenario with LOLP = 0.01% and VOLL $10 USD/kWh.

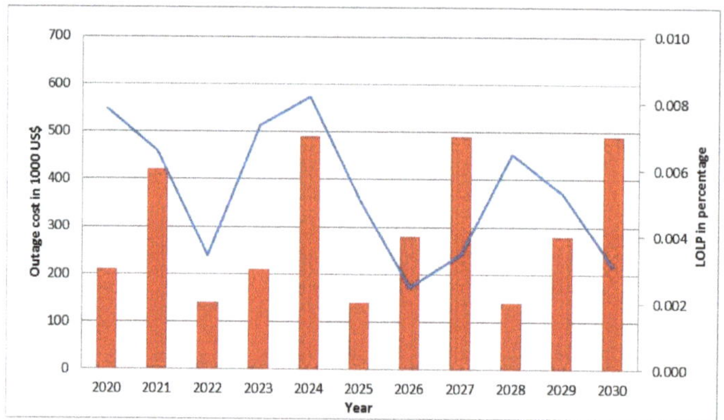

Figure 10. LOLP and OC in a scenario with LOLP = 0.01% and VOLL $10 USD/kWh.

The next group of scenarios was analyzed with the VOLL $2 USD/kWh for two cases, one without limiting the value of LOLP and one limiting the value of LOLP onto 1%. Load factors of power plant candidates remained almost the same on average, but the load factor of the 400 MW CCGT varied depending on the new thermal power plants put into operation, especially the coal-fired TPP. All the curves are shown in Figure 11.

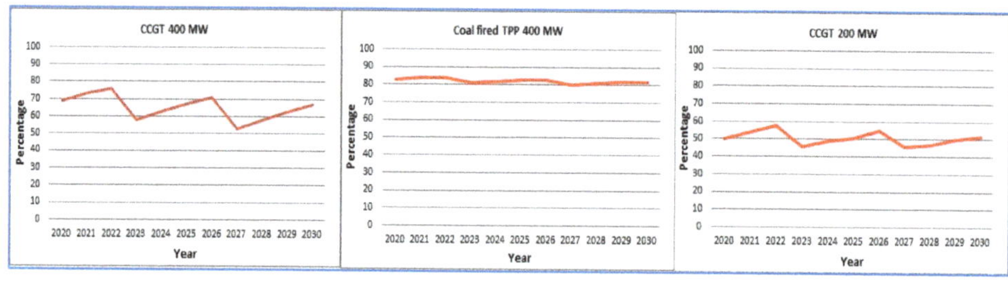

Figure 11. Load factor of new power plants in the LCP scenario and VOLL $2 USD/kWh.

The curves in Figure 12 convey the value of LOLP and the outage cost per year, which showed characteristic behaviors. One of those was year 2022, the fourth year when no TPPs were put into operation. The maximum achieved value of LOLP was 3.95%.

Load factors of new thermal power plants, especially the coal-fired TPP and the 200 MW CCGT, which were put into operation in the observed period, were relatively constant, whereby the load factor of the CCGT with higher capacity slightly decreased towards the end of the planned period, as seen in Figure 13.

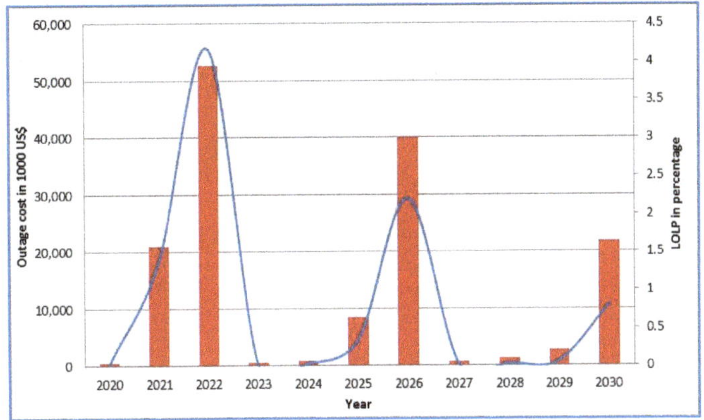

Figure 12. LOLP and OC in a scenario with LOLP = 0.01% and VOLL $2 USD/kWh.

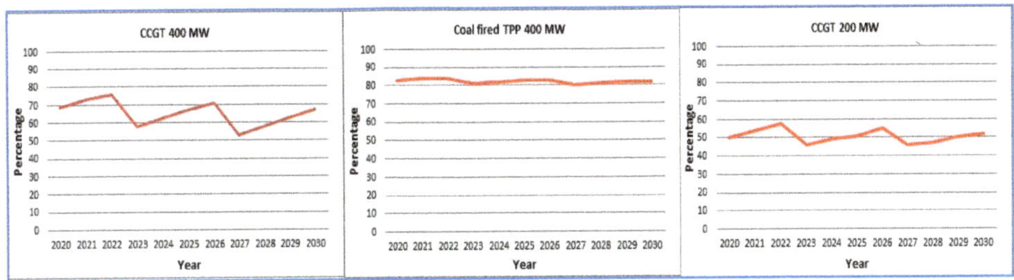

Figure 13. Load factor of new power plants in a scenario with LOLP = 1% and VOLL $2 USD/kWh.

The curves of the outage cost as well as the values of LOLP per year are presented in Figure 14. Characteristic years for this group of scenarios were 2021 and 2025 as well as the last year of the planned period.

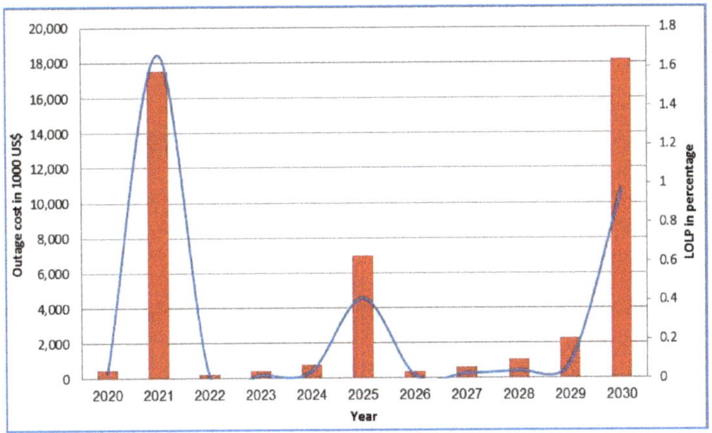

Figure 14. LOLP and OC in a scenario with LOLP = 1% and VOLL $2 USD/kWh.

5. Scenarios Comparison

The comparison of these four scenarios was possible based on three elements. The first comparison was of scenarios in the group with the same level of VOLL for two different levels of LOLP. The second comparison of scenarios was on the basis of the same level of LOLP for two levels of the VOLL. The third comparison was for different levels of VOLL and LOLP.

The comparison of scenarios with VOLL $10 USD/kWh and for LOLP of 0.01% without limitations showed significant differences in the second half of the observed period. If the scenarios were compared according to total installed power or the necessary new installed capacity, as shown in Figure 15, it was obvious that by 2019, the installed total power was almost the same, i.e., the necessary new installed capacity. As previously stated, the scenario without limitations corresponded to the LCP scenario, whereas the second scenario with LOLP 0.01% was the scenario with the optimized level of LOLP.

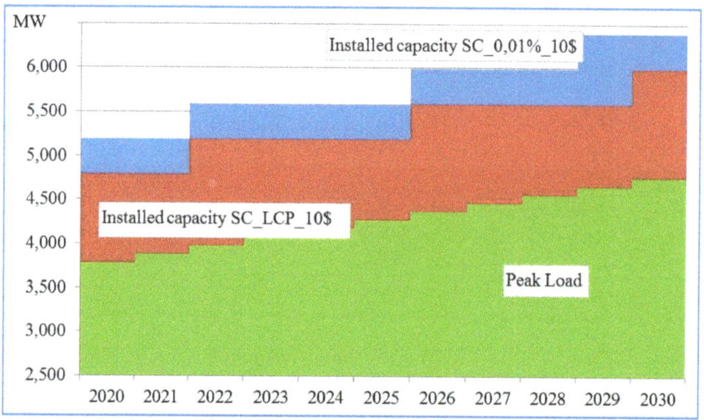

Figure 15. Scenario with VOLL $10 USD/kWh and LOLP = 0.01% versus the LCP scenario.

The comparison of scenarios with VOLL $2 USD/kWh and two levels of LOLP without limitations and with a limitation of 1% is shown in Figure 16. These two scenarios differed only in the dynamics of necessary realization in two years of the observed period. This was seen in the second half of the planned period in a way that one power plant was the candidate in the scenario with LOLP of $2 USD/kWh and was put into operation a year earlier than in the scenario without the limitation of LOLP. That meant that, in the group of scenarios with low outage costs, there was actually no difference between generation expansion planning using the method of LCP and the method of optimizing LOLP. The reason for this was that the influence or the share of costs referring to outage cost was almost negligible. These costs went from 0.3% for the scenario with LOLP 3% to 0.96% for the scenario without limitations of LOLP.

This would mean that the issue of customer security of supply for low levels of GDP was not relevant. That issue becomes relevant with the development of an economy and therefore the increase in total GDP, i.e., GDP per capita.

The comparison of two scenarios—in this case, the scenario with VOLL $2 USD/kWh and the level of LOLP without limitations compared to the scenario with VOLL $10 USD/kWh and the level of LOLP of 0.01%—regarding the necessary new construction of power plants is shown in Figure 16.

In this example, there were quantified differences between the necessary constructions of the generation facilities for the LCP scenario compared to the optimized level of the LOLP scenario. The conclusion was drawn that, for a developed economy measured by the VOLL $10 USD/kWh, it was necessary to dispose of the capacity in power plants higher than 400 to 800 MW. It should be noted that additional installed generation capacity was not necessary because, in both scenarios, it was modeled in the same way. However, it was necessary for fulfilling the criteria of customer security of supply. Namely, a high level of economic development consequently had a high level of outage cost, which then in turn necessitated a high level of customer security of supply, requiring additional necessary installed capacity for an equal level of electricity consumption. The difference between these two scenarios was critical. The first scenario corresponded to the method of LCP of the power system, whereby the second corresponded to the scenario with given criterion of LOLP, i.e., directly to the set level of customer security of supply. Apart from that, the difference was in the level of VOLL.

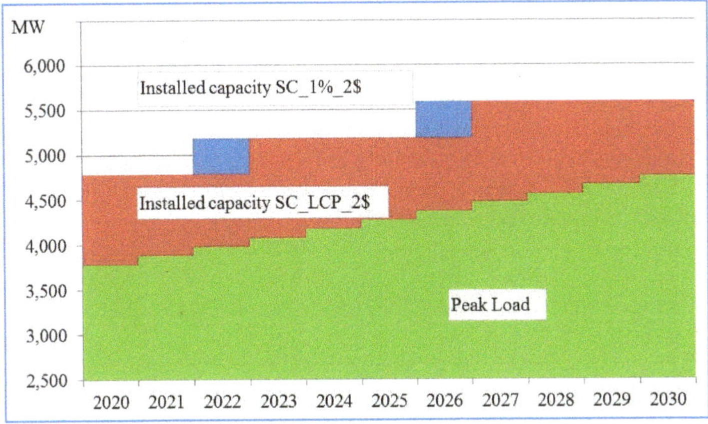

Figure 16. Scenario with VOLL $2 USD/kWh and LOLP = 1% versus the scenario with LCP.

Figure 17 shows the comparison of the necessary new construction of generation capacities for the scenarios corresponding to the LCP scenario for VOLL $2 USD/kWh and $10 USD/kWh. The differences arose from the contribution of outage cost.

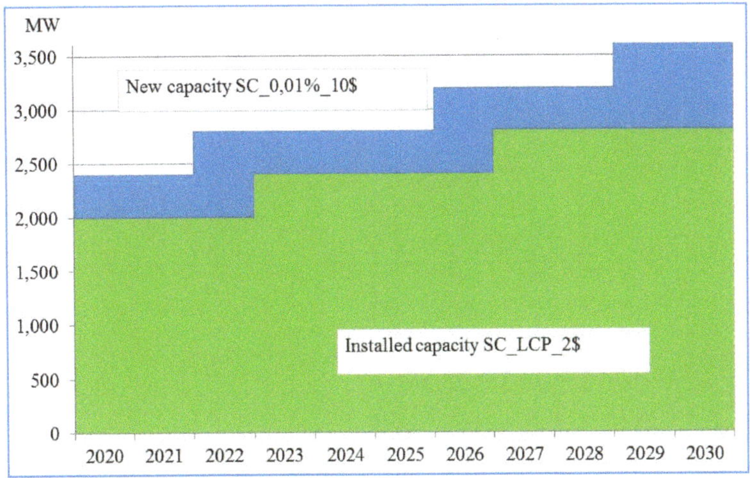

Figure 17. Scenario with VOLL $10 USD/kWh and LOLP = 0.01% versus the scenario with VOLL = $2 USD/kWh and LCP.

6. Conclusions

This paper presented a developed method for incorporating reliability criterion in the optimization of power system expansion planning. Analysis of security of supply was significant when power systems were poorly interconnected with other power systems, which is once again significant when considering the regional electricity market as a real business environment.

The basic measure of security of supply is the reserve capacity in the power system. The better developed an economy is, the more justified a bigger reserve capacity is because of a higher value of VOLL, and vice versa. This means that security of supply is an economic category with a different value in economies on different levels of development. The optimal level of LOLP differs for different levels of VOLL as well as analyzed technology of new power plants, and this was clearly presented in the paper. It was shown that LOLP should be calculated rather than proposed as a pre-specified value, since LOLP depends on VOLL, which is a certain measure of economic development and is changing through time in undeveloped economies. The key variable defining the necessary reserve capacity in a power system is VOLL, whereas in developed economies, the key variable is LOLP. Therefore, LOLP has a different meaning in different circumstances. In the case of LCP, LOLP is an economic parameter, and in the case of optimizing, LOLP is a technical parameter.

Self-sufficiency of a power system cannot be economically justified, since in the case of a big reserve capacity, the load factor of power plants decreases, thus the economics of such power plants are questionable. Power systems with a big reserve capacity need a regional electricity market to improve the economics of such power plants. Therefore, it is reasonable to expect that electricity generation capacities would be economical if they were on an electricity market, which would have a favorable influence on the price of electricity at the market. As was shown in the paper, load factor decreases in the case of a small value of LOLP, and this is not acceptable from an economical point of view.

One should note that electricity trade between developed and undeveloped countries would put undeveloped countries at a disadvantage. The reason lies in the fact that developed countries will have the economic power to construct enough (or more) generation capacities for the appropriate LOLP, whereby undeveloped countries will not be able to do so, since a significantly lower LOLP is optimal for them. LOLP as main criteria for security of supply depends on VOLL, and characteristics of power plants technology are changing over time. The main conclusion of the paper is that, for generation expansion planning on the electricity market, the least acceptable LOLP rather than the LCP should be a criterion for the necessary constructions in a power system; this paper particularly referred to the generation capacity needed to fit the economic condition of the considered country and its power system.

Author Contributions: Conceptualization, Formal Analysis, Writing—Original Draft Preparation, G.S., Methodology and Validation, M.Z., Supervision, Writing-Review & Editing D.Š.

Funding: This research received no external funding.

Conflicts of Interest: The authors declare no conflict of interest.

References

1. Directive 2005/89/EC of the European Parliament and of the Council. 2006. Available online: https://eur-lex.europa.eu/legal-content/EN/TXT/PDF/?uri=CELEX:32005L0089&rid=7 (accessed on 15 March 2019).
2. ENTSO-e Target Methodology for Adequacy Assessment, October 2014. Available online: https://docstore.entsoe.eu/Documents/SDC%20documents/SOAF/ENTSO-E_Target_Methodology_for_Adequacy_Assessment.pdf (accessed on 15 March 2019).
3. Identification of Appropriate Generation and System Adequacy Standards for the Internal Electricity Market, European Commission, Directorate-General for Energy, Directorate B—Internal Energy Market, Unit B2—Wholesale markets; electricity & gas, Brussels. 2016. Available online: https://ec.europa.eu/energy/sites/ener/files/documents/Generation%20adequacy%20Final%20Report_for%20publication.pdf (accessed on 15 March 2019).

4. Ovaere, M.; Heylen, E.; Proost, S.; Deconinck, G.; Van Hertem, D. How detailed value of lost load data impact power system reliability decisions - a trade-off between efficiency and equity, SSRN Electronic Journal, January 2016. Available online: https://poseidon01.ssrn.com/delivery.php?ID=565072116003087011122004124023073067058053071002044010118007113106095012108104116070110021048011118001020023030069113027022000118026088041040106015004030068103064003040085110000092008096096072065100011067099027122121080125000116122112082079083001077&EXT=pdf (accessed on 15 March 2019).
5. Electric Grid Expansion Planning with High Levels of Variable Geberation, Oak Ridge Laboratory, February 2016. Available online: https://info.ornl.gov/sites/publications/files/Pub59037.pdf (accessed on 15 March 2019).
6. Estimating the Economically Optimal Planning Reserve Margin, Energy and Environmental Economics, Inc., USA. 2015. Available online: https://www.epelectric.com/files/html/PRM_Report.pdf (accessed on 15 March 2019).
7. Hladik, D.; Fraunholz, C.; Manz, P.; Kühnbach, M.; Kunze, R. A Multi-Model Approach to Investigate Security of Supply in the German Electricity Market. In Proceedings of the 2018 15th International Conference on the European Energy Market (EEM), Lodz, Poland, 27–29 June 2018; pp. 1–5.
8. Sadeghi, A.; Torbaghan, S.S.; Gibescu, M. Benefits of Clearing Capacity Markets in Short Term Horizon: The Case of Germany. In Proceedings of the 2018 15th International Conference on the European Energy Market (EEM), Lodz, Poland, 27–29 June 2018; pp. 1–5.
9. Tasios, N.; Capros, P.; Zampara, M. Model-based analysis of possible capacity mechanisms until 2030 in the European internal electricity market. In Proceedings of the 11th International Conference on the European Energy Market (EEM14), Krakow, Poland, 28–30 May 2014; pp. 1–5.
10. Chao, H.; Lawrence, D.J. How capacity markets address resource adequacy. In Proceedings of the 2009 IEEE Power & Energy Society General Meeting, Calgary, AB, Canada, 26–30 July 2009; pp. 1–4.
11. Łyżwa, W.; Mielczarski, W. In Proceedings of Meeting criteria for capacity markets. In Proceedings of the 11th International Conference on the European Energy Market (EEM14), Krakow, Poland, 28–30 May 2014; pp. 1–6.
12. International Atomic Energy Agency: Expansion Planning for Electrical Generating Systems, A Guidebook, Technical reports series No. 241, Vienna, 1984. Available online: https://www-pub.iaea.org/MTCD/publications/PDF/TRS1/TRS241_Web.pdf (accessed on 15 March 2019).
13. International Atomic Energy Agency: Energy and Nuclear Power Planning in Developing Countries, Technical reports series No. 245, Vienna, 1985. Available online: https://www-pub.iaea.org/MTCD/publications/PDF/trs1/trs245_web.pdf (accessed on 15 March 2019).
14. International Atomic Energy Agency: Wien Automatic system Planning package (WASP), A computer code for power generation system expansion planning, Volume I and II, Vienna, 1994. Available online: https://www-pub.iaea.org/MTCD/Publications/PDF/CMS-16.pdf (accessed on 15 March 2019).
15. Alferidi, A.; Karki, R. Development of Probabilistic Reliability Models of Photovoltaic System Topologies for System Adequacy Evaluation Title of Site, MDPI. *Appl. Sci.* **2017**, *7*, 176. [CrossRef]
16. Curtin, J.; Doherty, T. *The Value of Lost Load, the Market Price Cap and the Market Price Floor-A Consultation Paper, All Island Project*; SEM Committee: Belfast, UK, 2007.
17. Cambridge Economic Policy Associates Ltd, Study on the estimation of the Value of Lost Load of electricity supply in Europe. 2018. Available online: https://www.acer.europa.eu/en/Electricity/Infrastructure_and_network%20development/Infrastructure/Documents/CEPA%20study%20on%20the%20Value%20of%20Lost%20Load%20in%20the%20electricity%20supply.pdf (accessed on 15 March 2019).
18. Eurelectric, Power Statistics & Trends 2015. Available online: https://www.eurelectric.org/media/1992/power-statistics-and-trends-the-five-dimentions-of-the-energy-union-lr-2015-030-0641-01-e.pdf (accessed on 15 March 2019).
19. VGB Powertech: Investment and Operation Cost Figures—Generation Portfolio, November 2011. Available online: https://www.vgb.org/vgbmultimedia/LCOE_Final_version_status_09_2012-p-5414.pdf (accessed on 15 March 2019).
20. Mott MacDonald: UK Electricity Generation Cost Update, Brighton, UK. 2010. Available online: https://assets.publishing.service.gov.uk/government/uploads/system/uploads/attachment_data/file/65716/71-uk-electricity-generation-costs-update-.pdf (accessed on 15 March 2019).
21. Cramton, P.; Stoft, S. A Capacity Market that Makes Sense. *Electri. J.* **2005**, *18*, 43–54. [CrossRef]

22. A. Creti, University of Toulouse, N. Fabra, Universidad Carlos II de Madrid, Capacity Market for Electricity, Center for the Study of Energy Markets (CSEM) Working Paper Series, University of California Energy Institute, Berkeley, California, USA. 2004. Available online: https://cloudfront.escholarship.org/dist/prd/content/qt4dd7f3mq/qt4dd7f3mq.pdf?t=kro6lg (accessed on 15 March 2019).
23. Hogan, W.W. Connecting Reliability Standards and Electricity Markets, Mossavar-Rahmani Center for Business and Government John F. Kennedy School of Government Harvard University Cambridge, Massachusetts, Atlanta. 2005. Available online: https://sites.hks.harvard.edu/fs/whogan/Hogan_hepg_120805.pdf (accessed on 15 March 2019).

MDPI
St. Alban-Anlage 66
4052 Basel
Switzerland
Tel. +41 61 683 77 34
Fax +41 61 302 89 18
www.mdpi.com

Energies Editorial Office
E-mail: energies@mdpi.com
www.mdpi.com/journal/energies

www.ingramcontent.com/pod-product-compliance
Lightning Source LLC
LaVergne TN
LVHW070906170726
843515LV00005B/712

9783036507729